TOBACCO BARNS

PHOTOGRAPHS BY LIGON B. FLYNN

BOOK DESIGN **CARL LORENZ** and **STEVEN McANALLY** with help from **Howard Garriss** and **Frank Huffman**
INTRODUCTION **CHERYL WILDER**

This book would not have been possible without the efforts and enthusiasm of Megan Johnson and James Flint. They were responsible for gathering the original negatives and producing most of the prints used here and for many other purposes. In addition to producing these materials for a museum exhibit in the spring of 2002 entitled "Down Tobacco Road" at the Cape Fear Museum in Wilmington, NC, Megan and James intended to produce a book via a grant for which they did much research and work not shown here. Thanks also to Ryland Fox for his help in the printing and presentation of these photographs.

This book available via print on demand at lulu.com

First Edition

All photographs are by the author.
ligonflynnarchitect.com

ISBN 978-0-6151-3981-4

Printed in the United States of America

INTRODUCTION

The vernacular variations of flue cured tobacco barns in southern Virginia, North Carolina, and South Carolina inspired years of stopping along the roadside to capture their diversity. Tobacco crops were established in Virginia before farmers realized the Carolinas provided ideal tobacco growing conditions. The change in terrain and climate as the land flattened in the Carolinas affected the size of crops, which in turn changed the placement of tobacco barns in the field themselves. In mountainous Virginia, barns were built on the edge of small crops next to neighboring woods which implemented natural shade for workers. Whereas, in the Carolinas, the flatlands provided farmers an ability to grow larger crops, therefore, barns were built centrally in the field for convenience. Carolina farmers continued adapting to their environment by changing the roof overhangs on tobacco barns to create shade for field workers. These few differences in fixed environment from state to state changed the architectural dynamics of tobacco barns, yet their functional similarities still distinguished them as unique from other farmhouse barns. And though the barns were built by farmers with no formal architectural training, the adaptation of functional detail found in tobacco barns defined these men as vernacular architects.

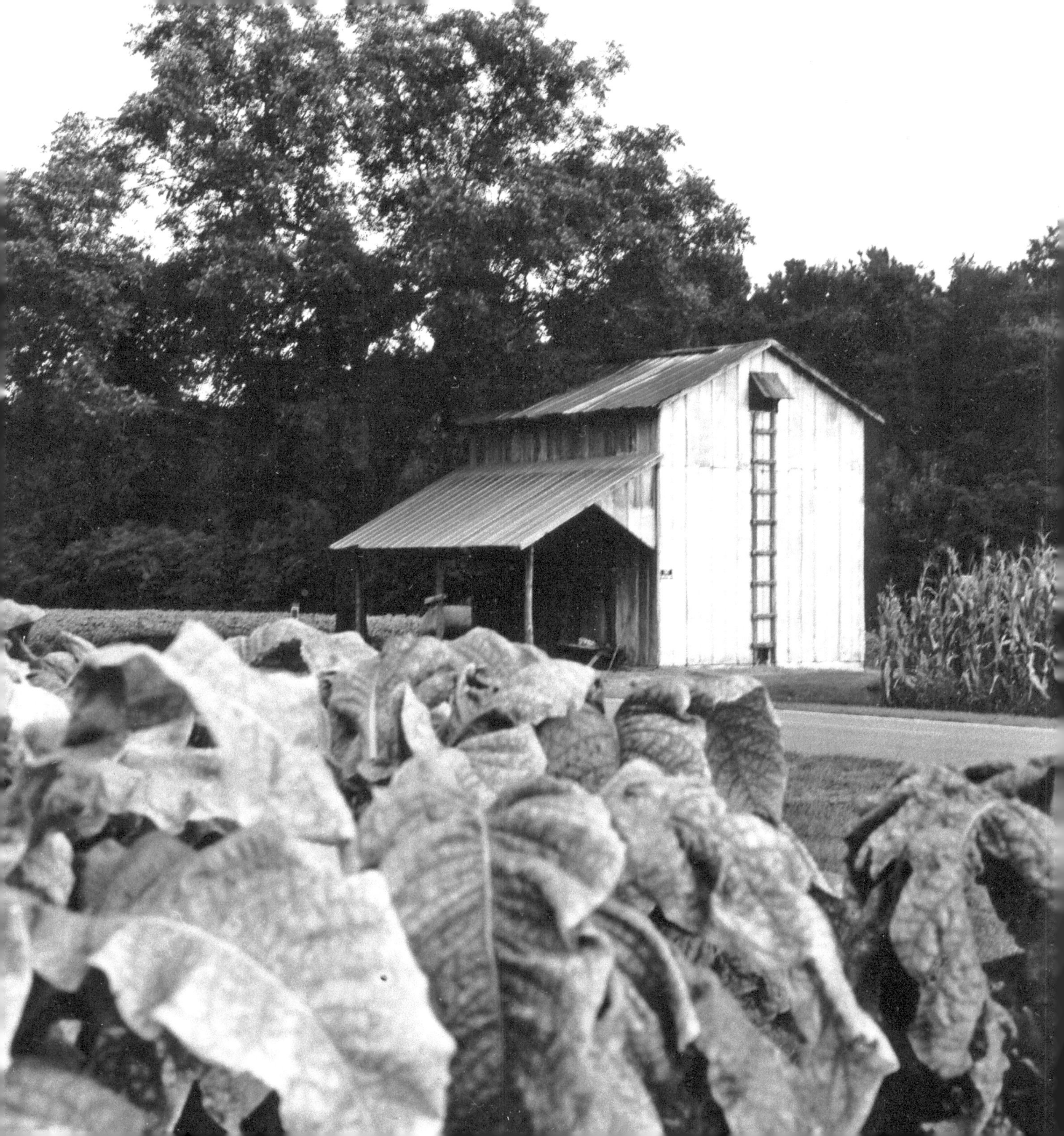

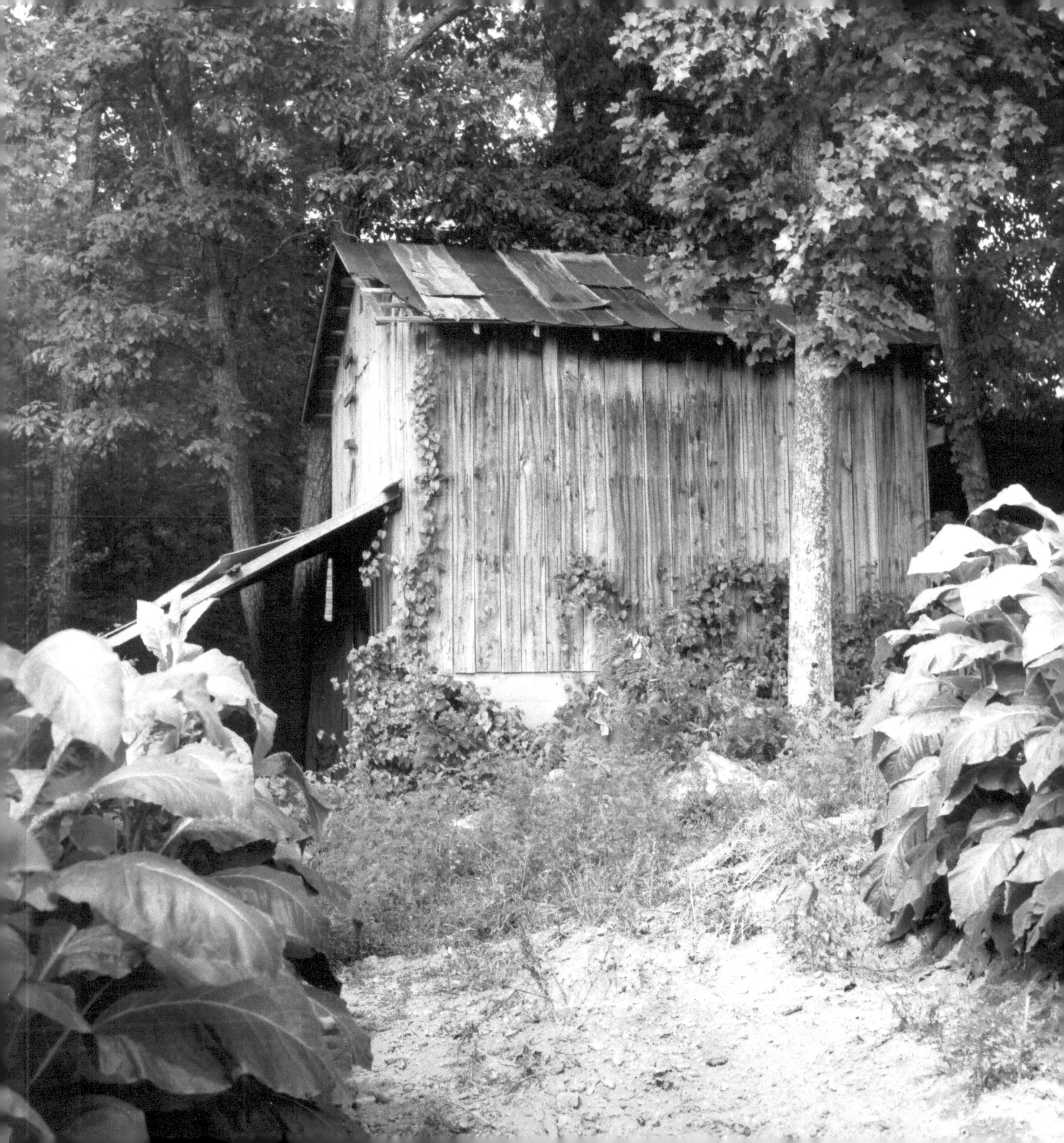

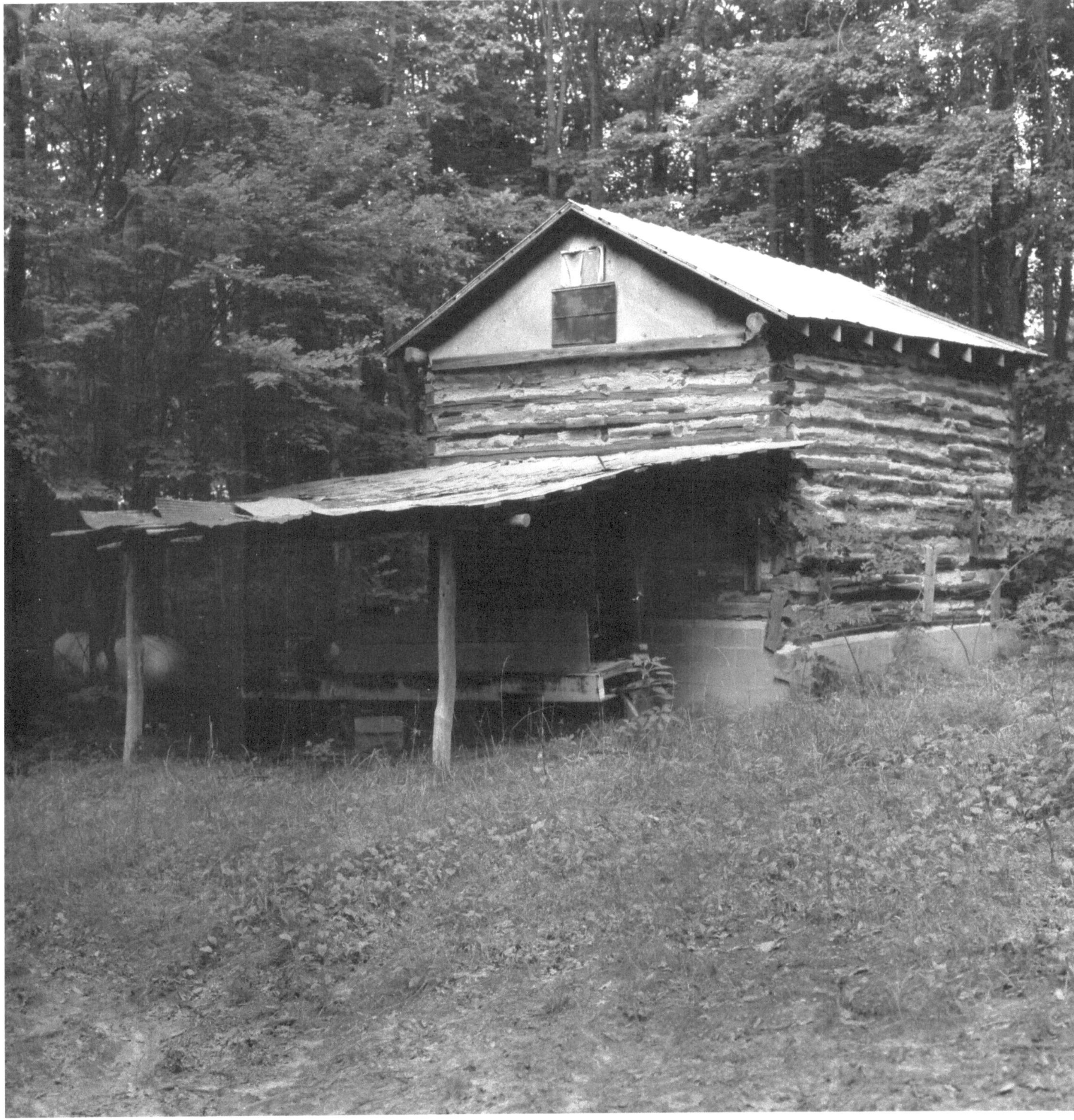

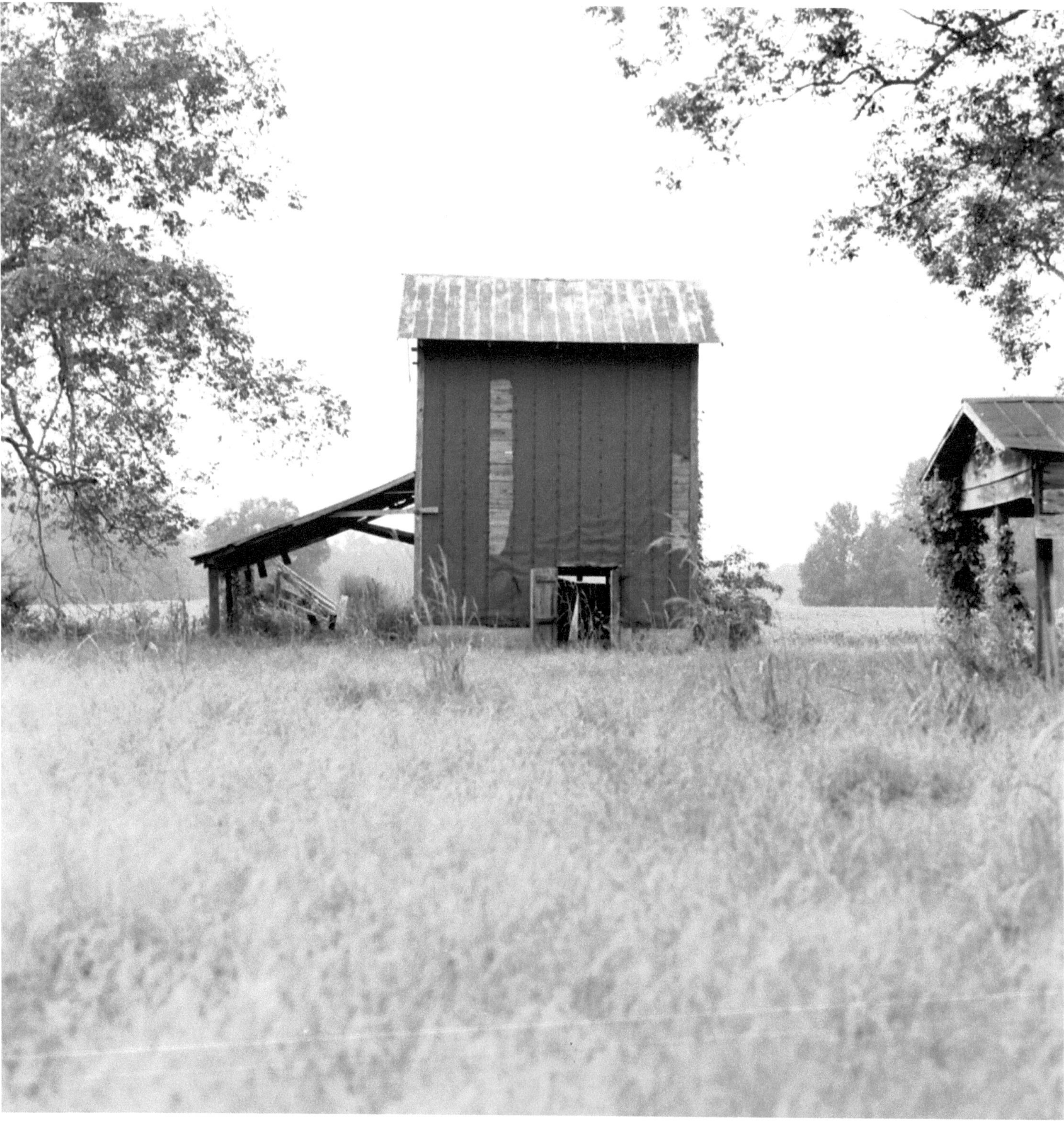

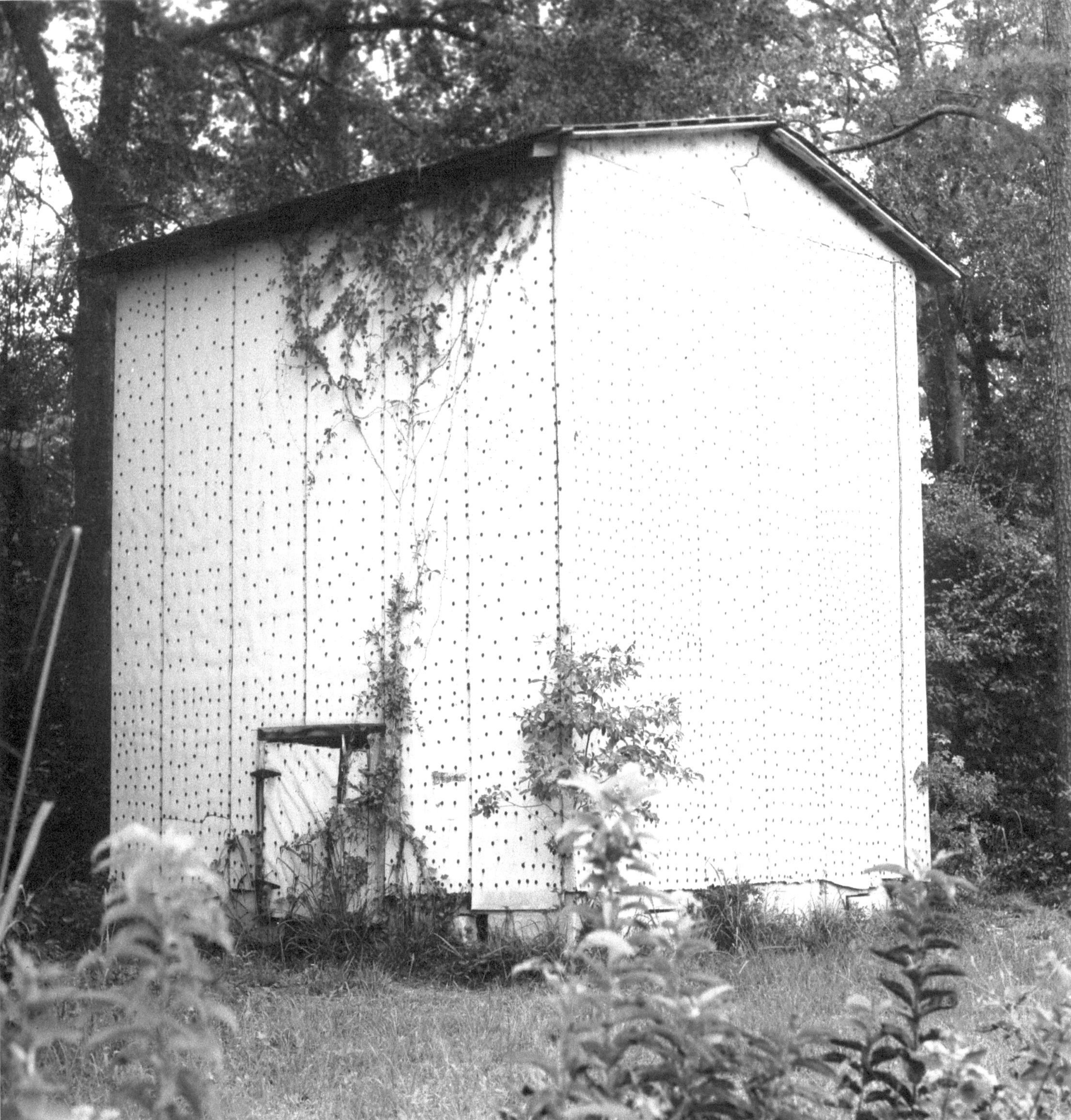

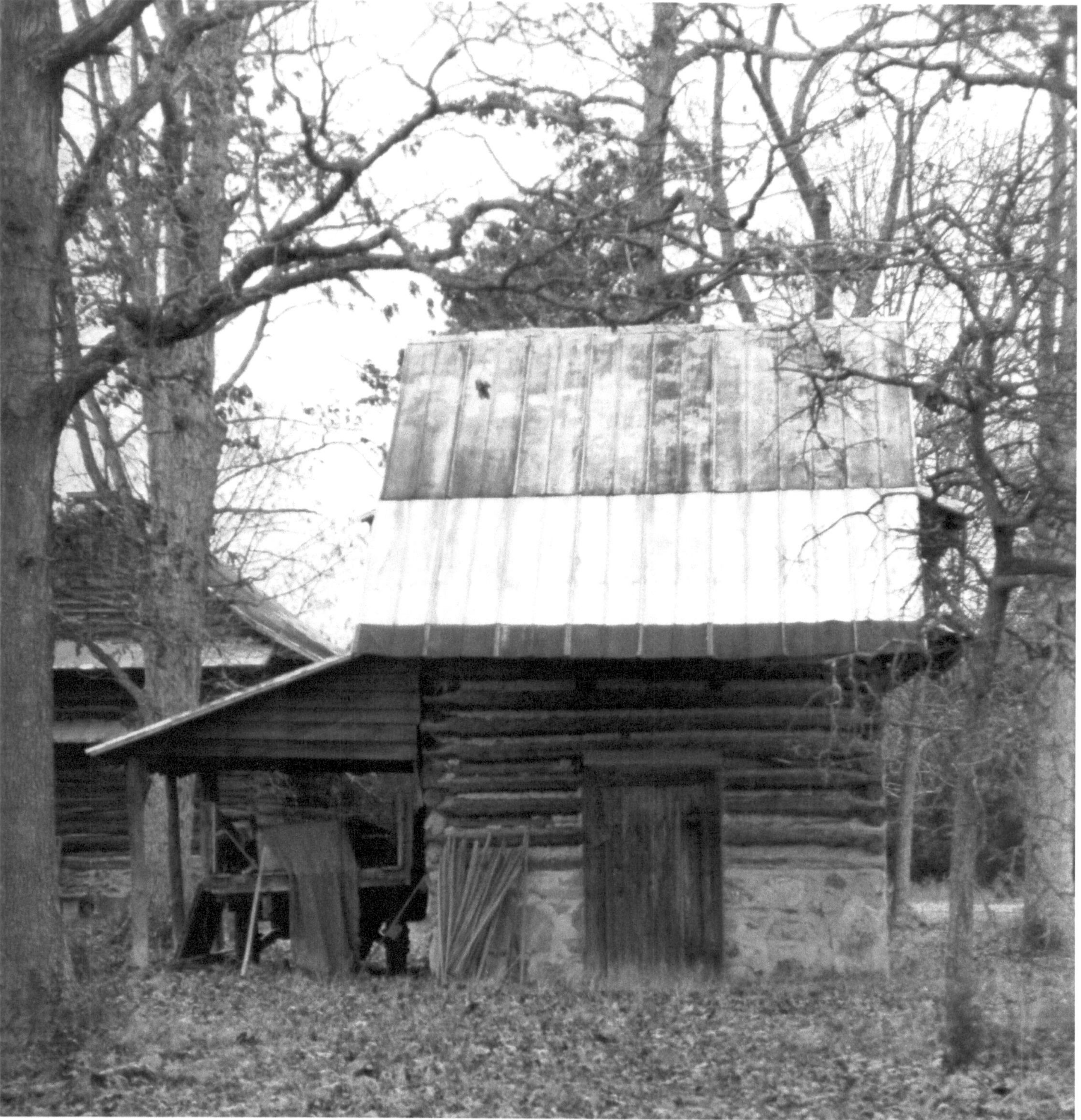

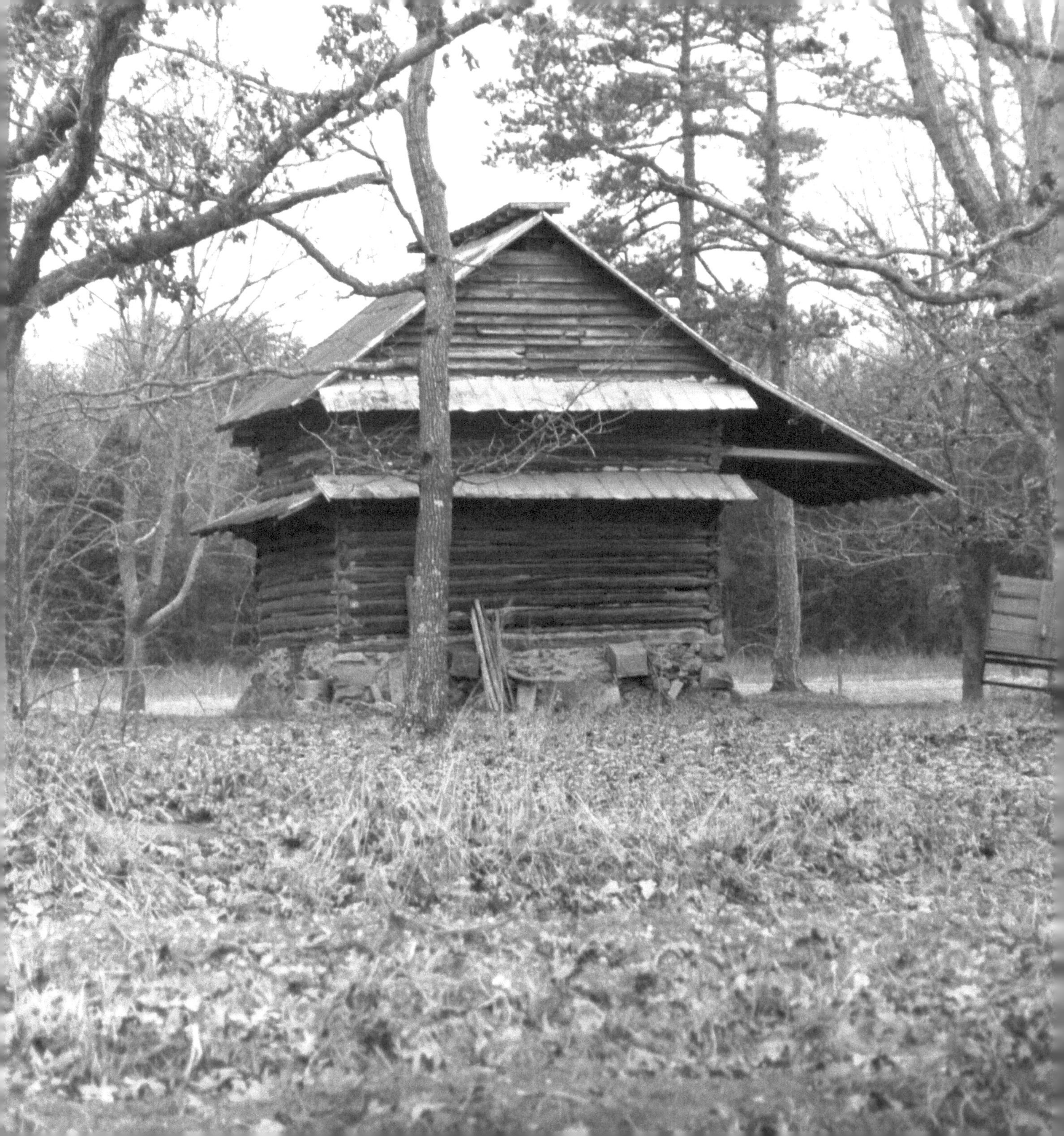

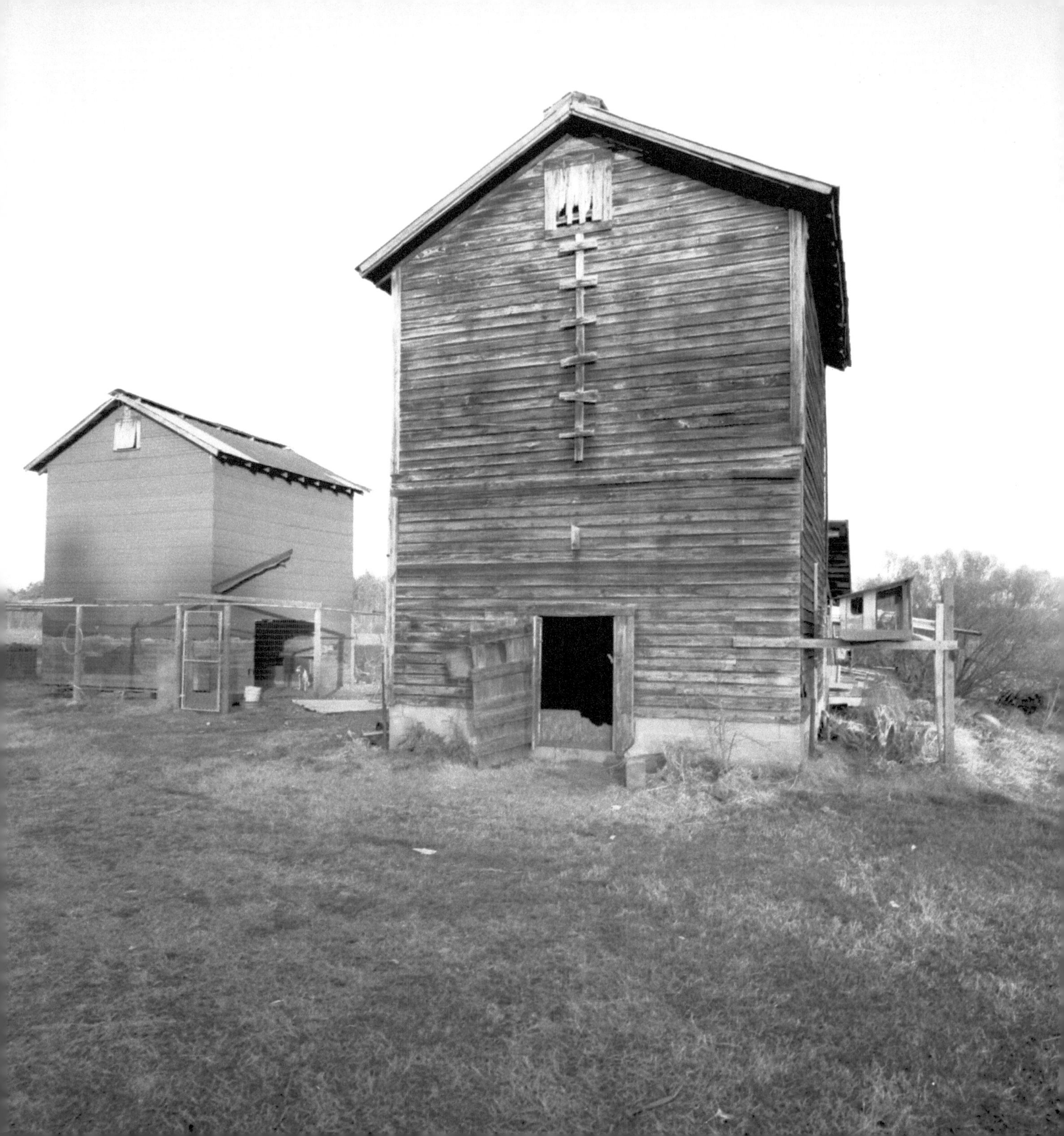

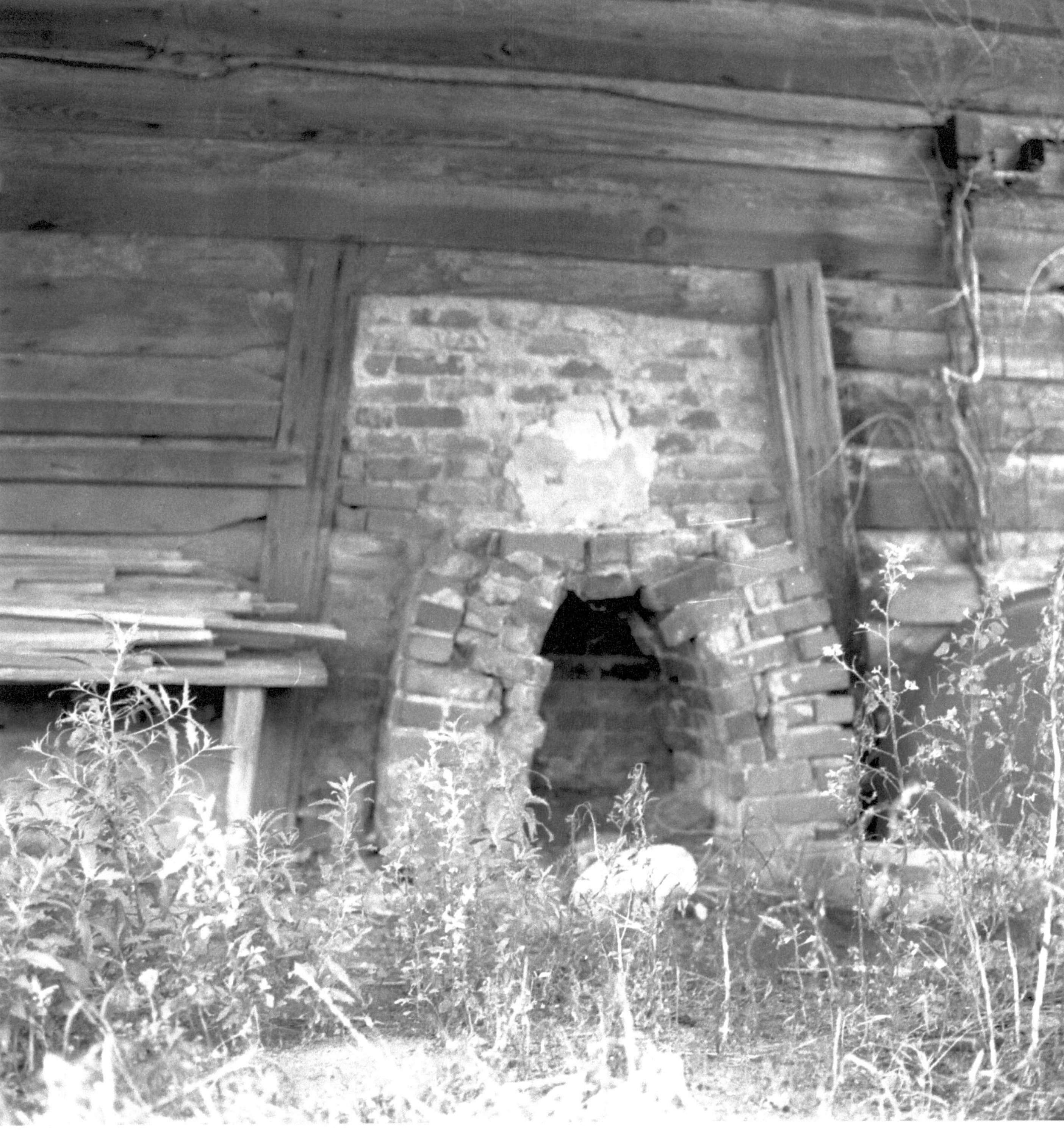

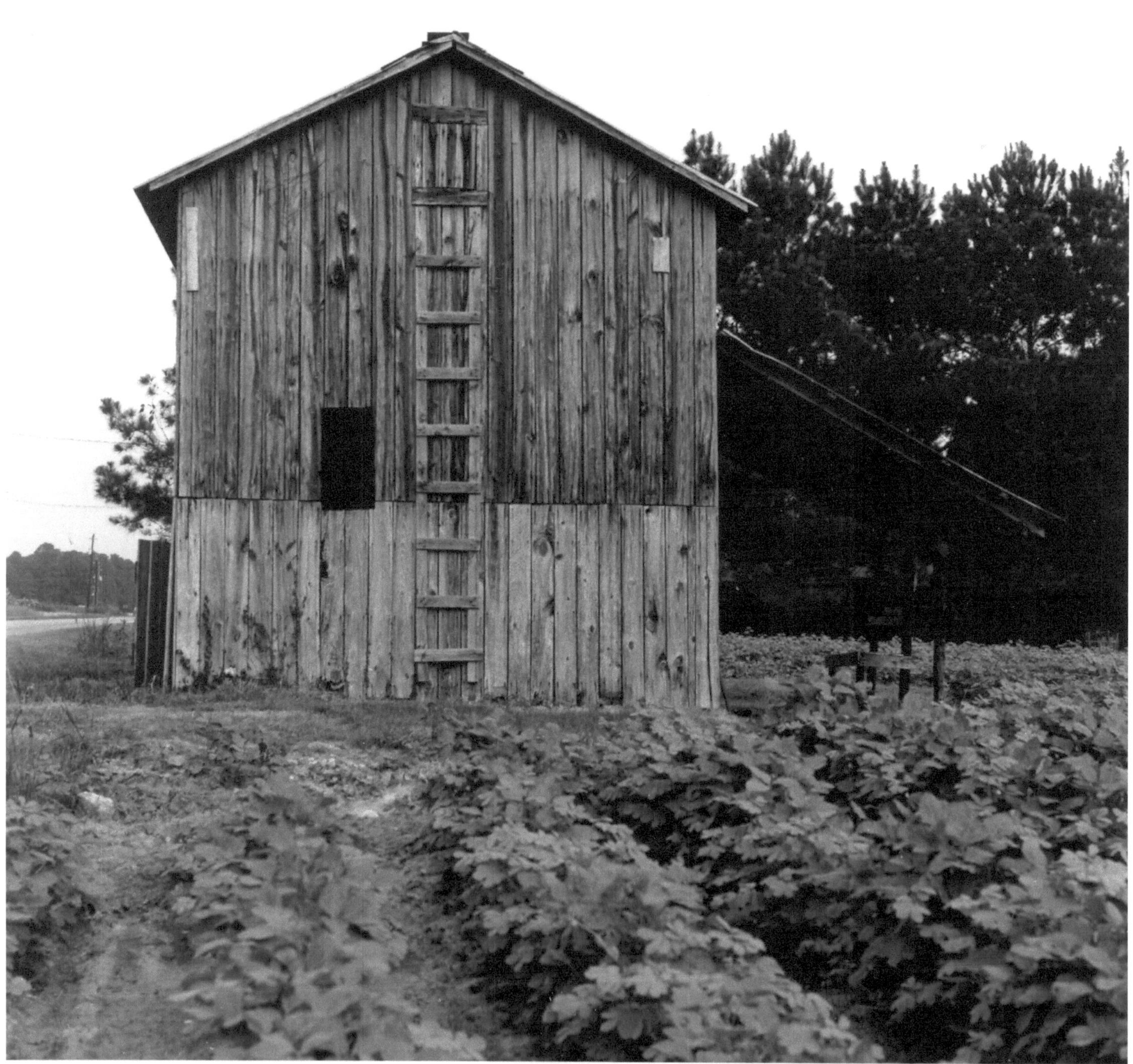

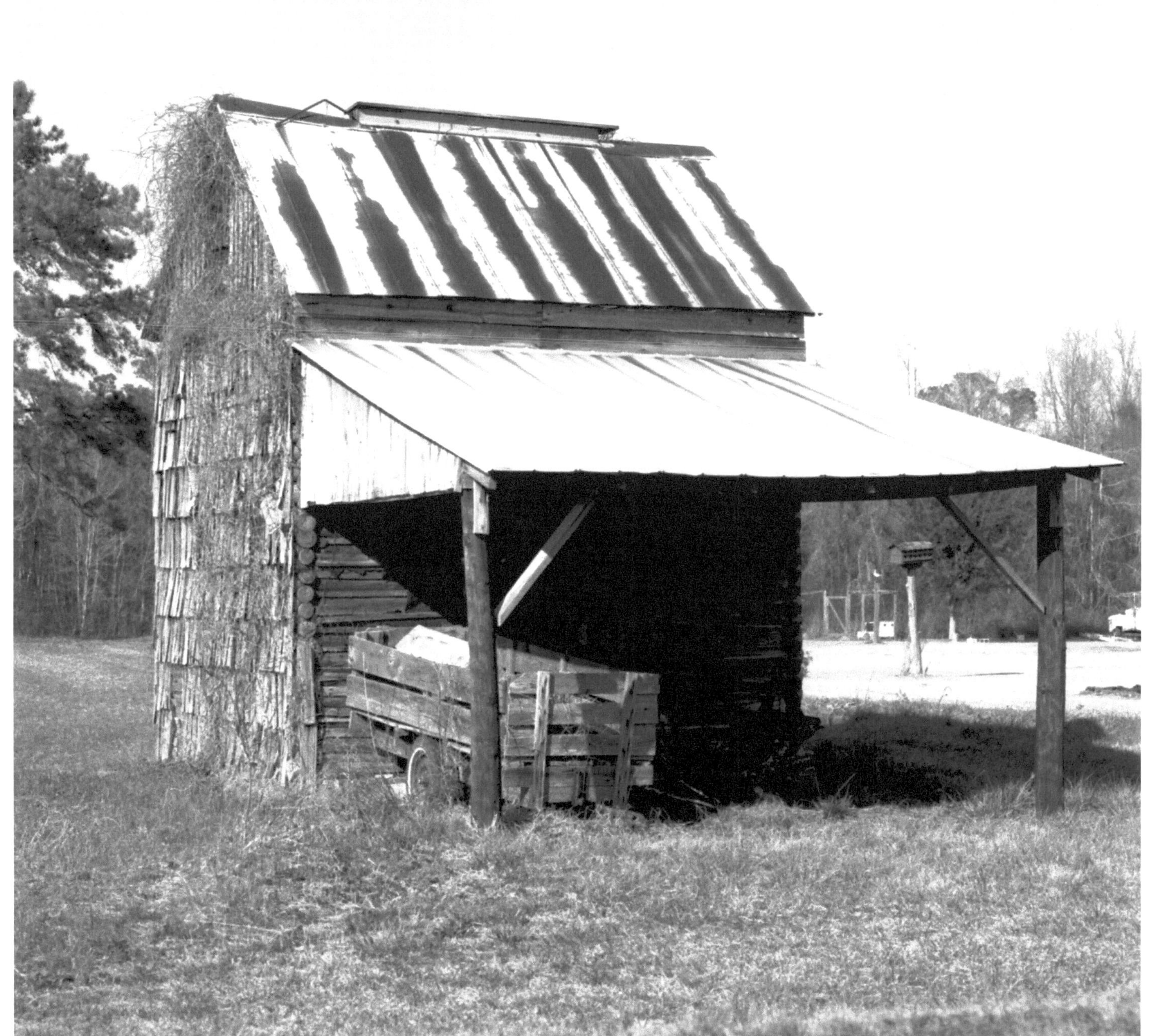

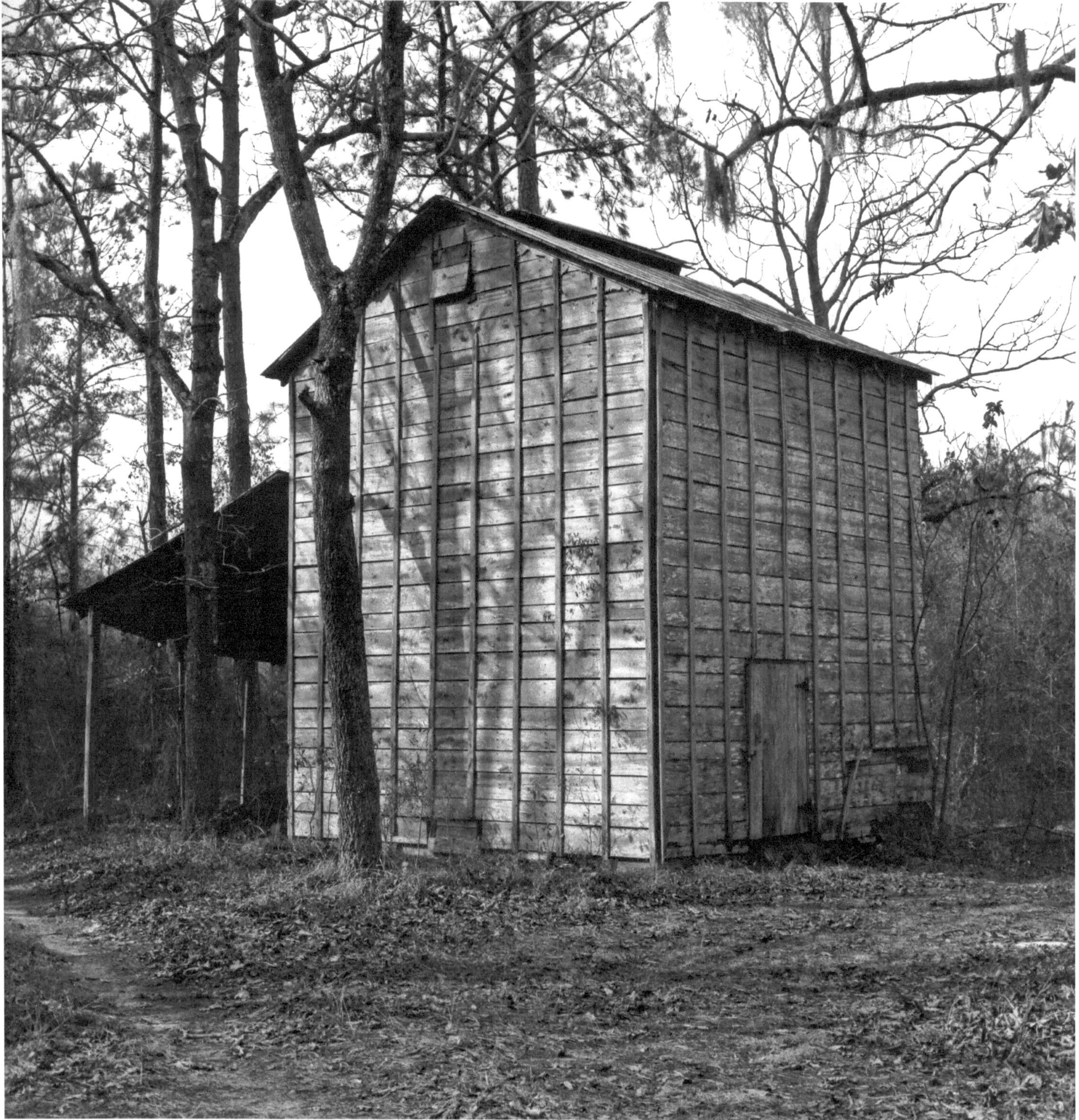

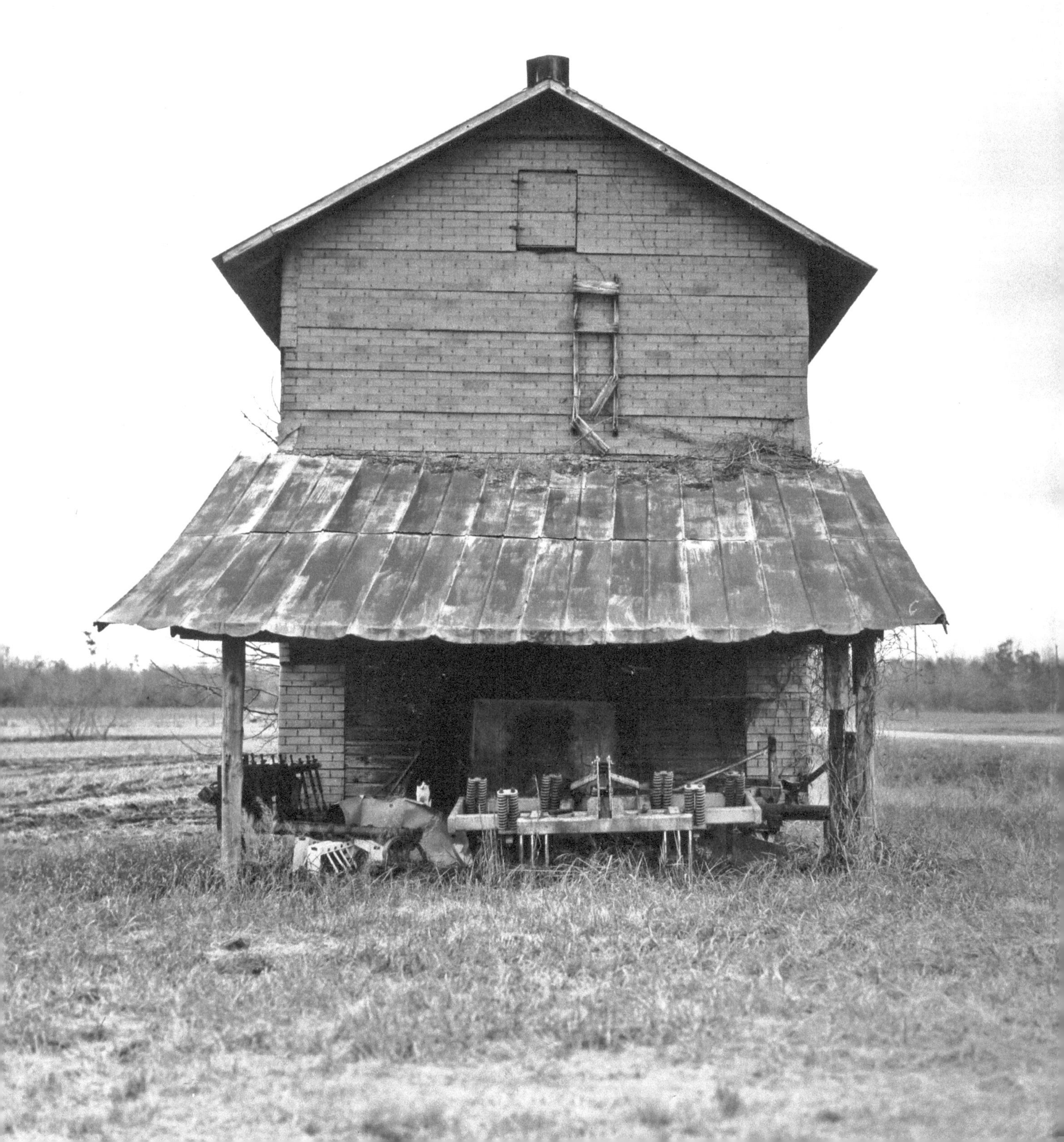

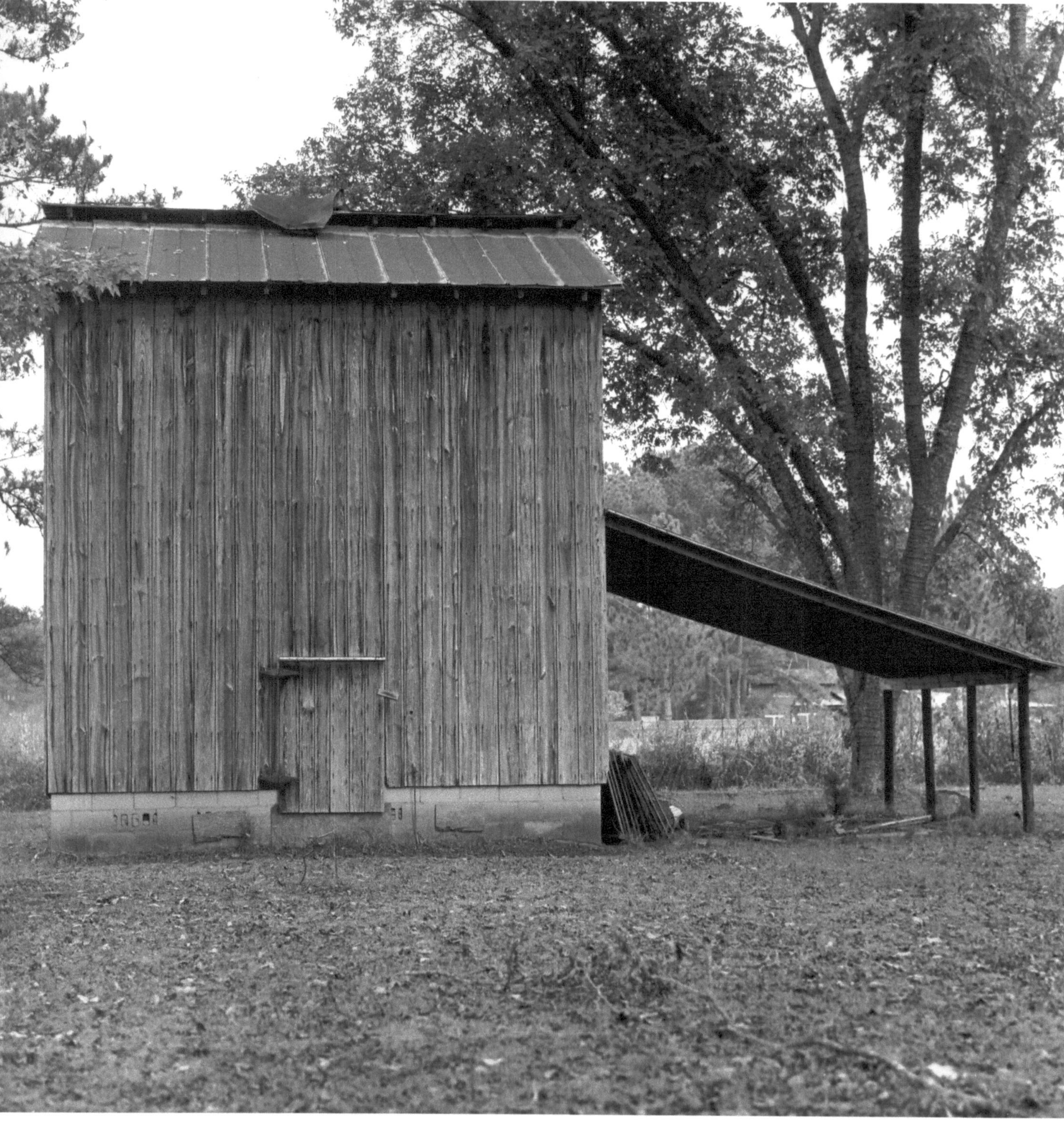

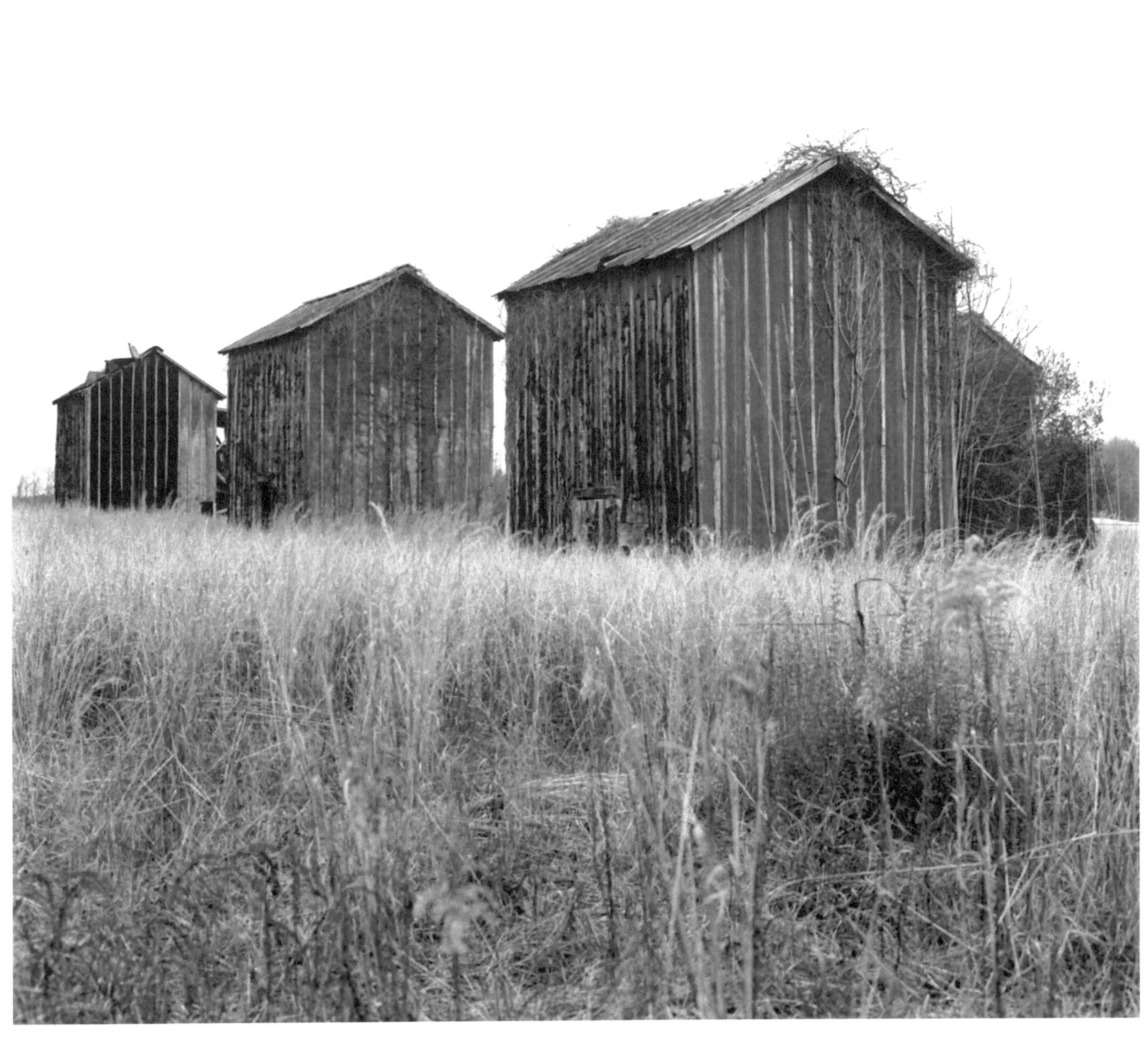

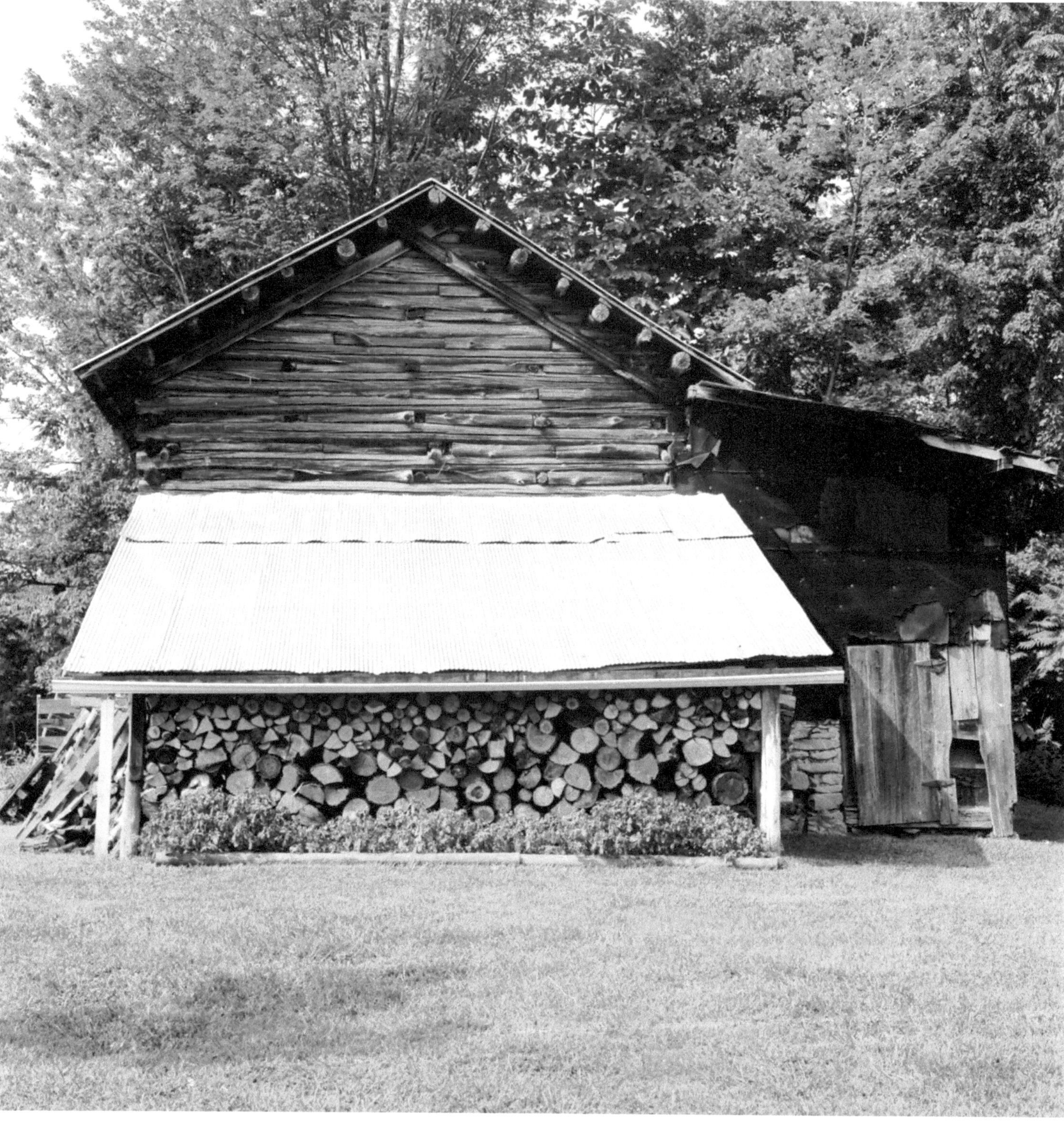

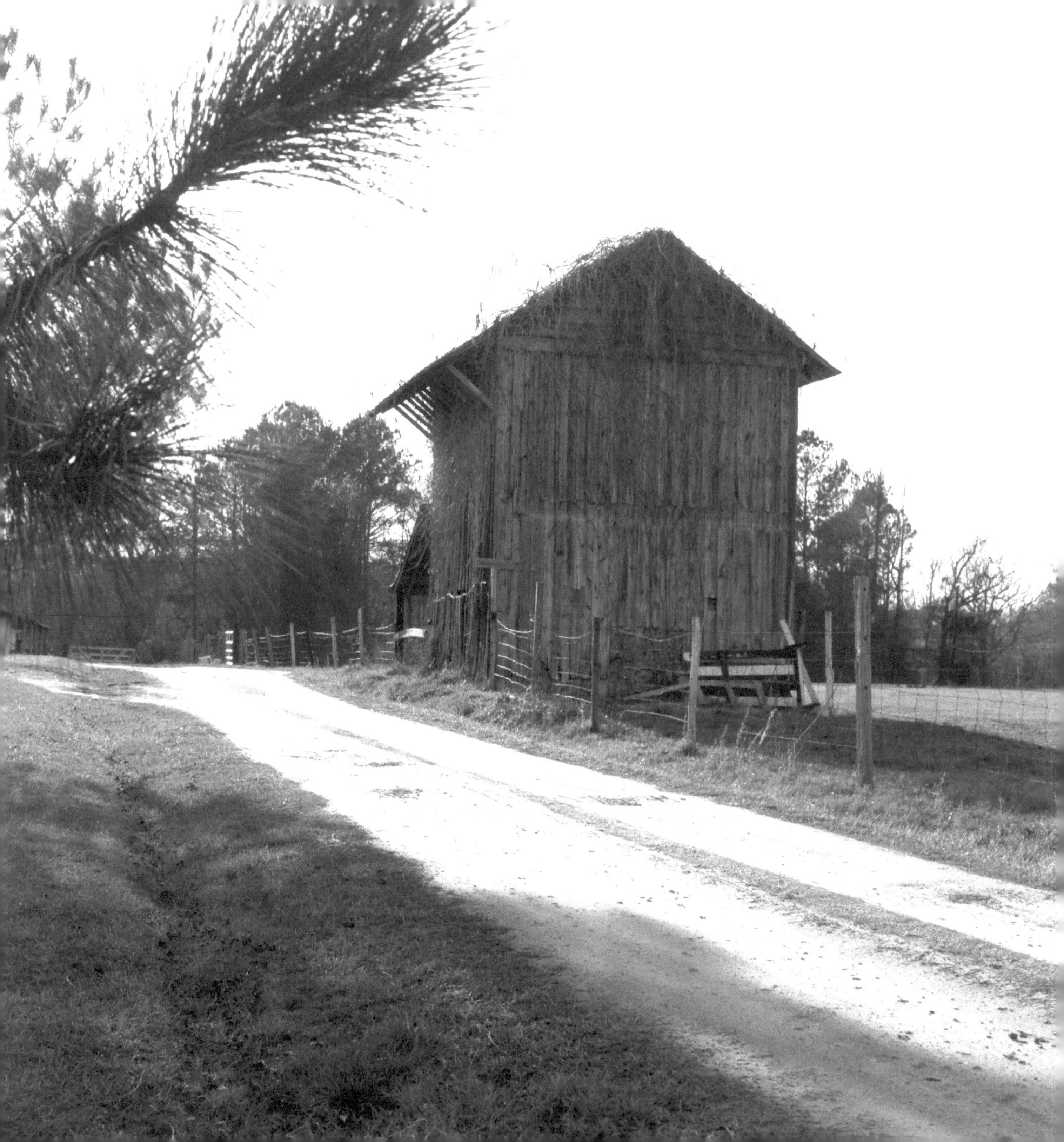

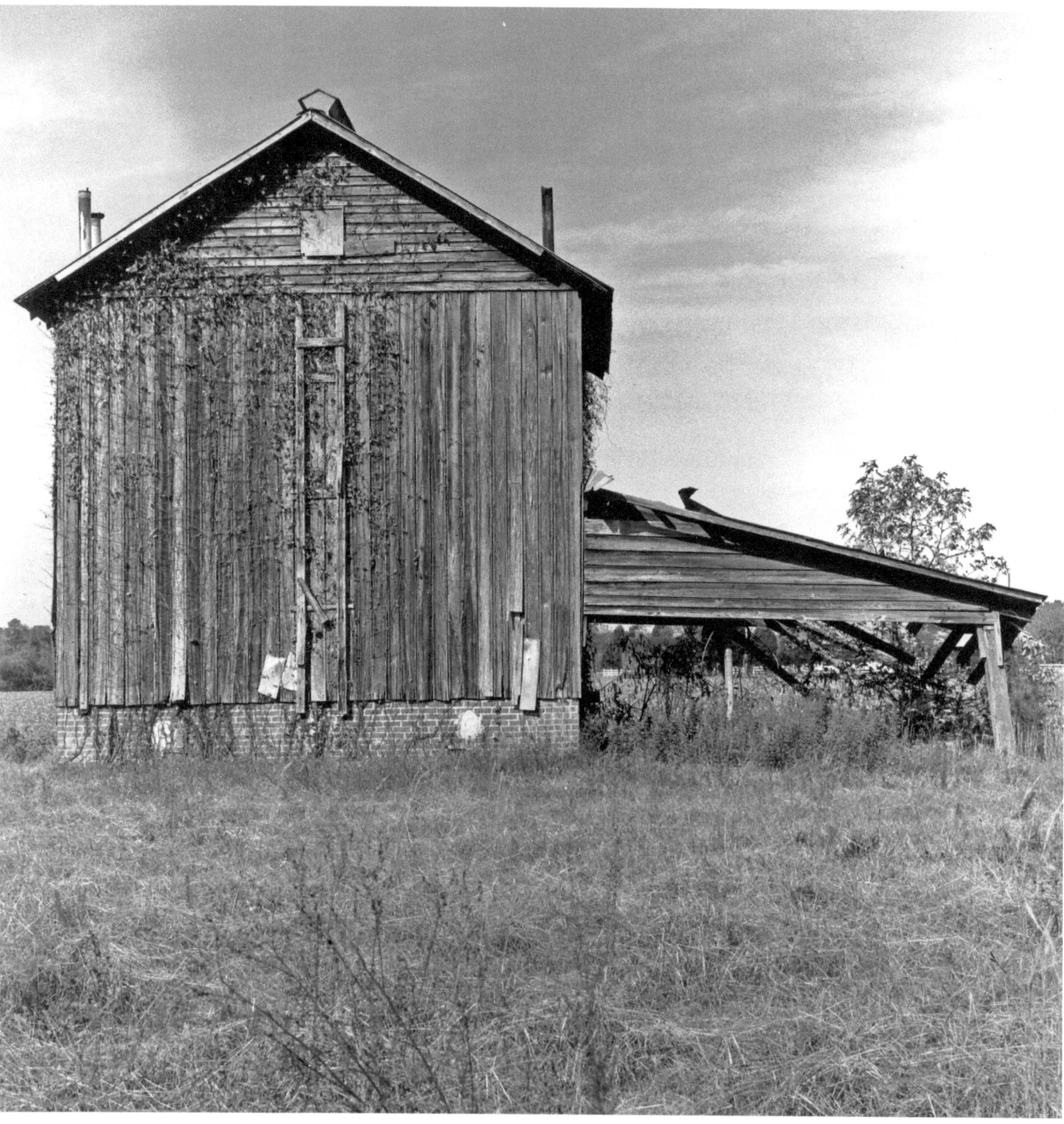

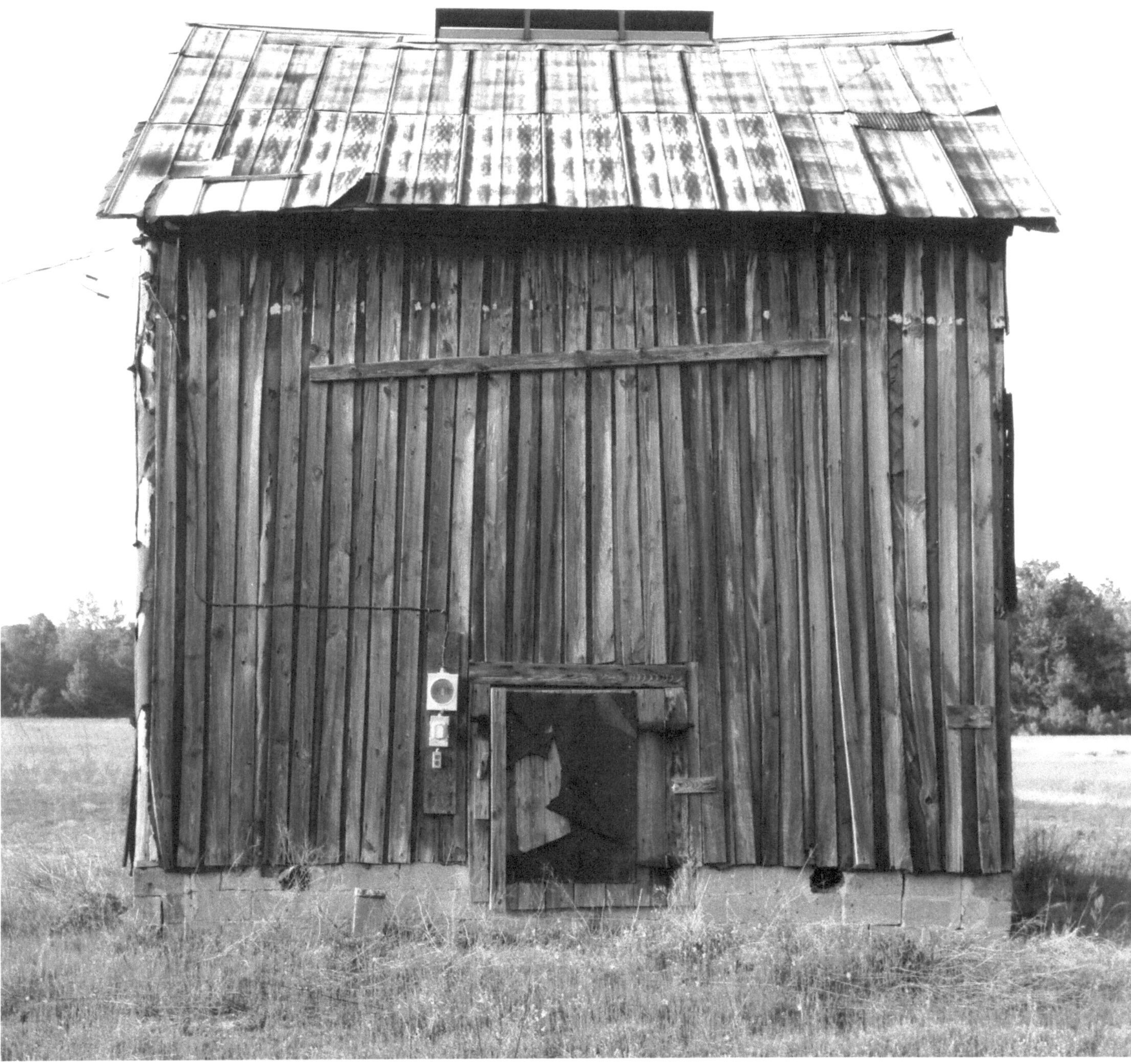

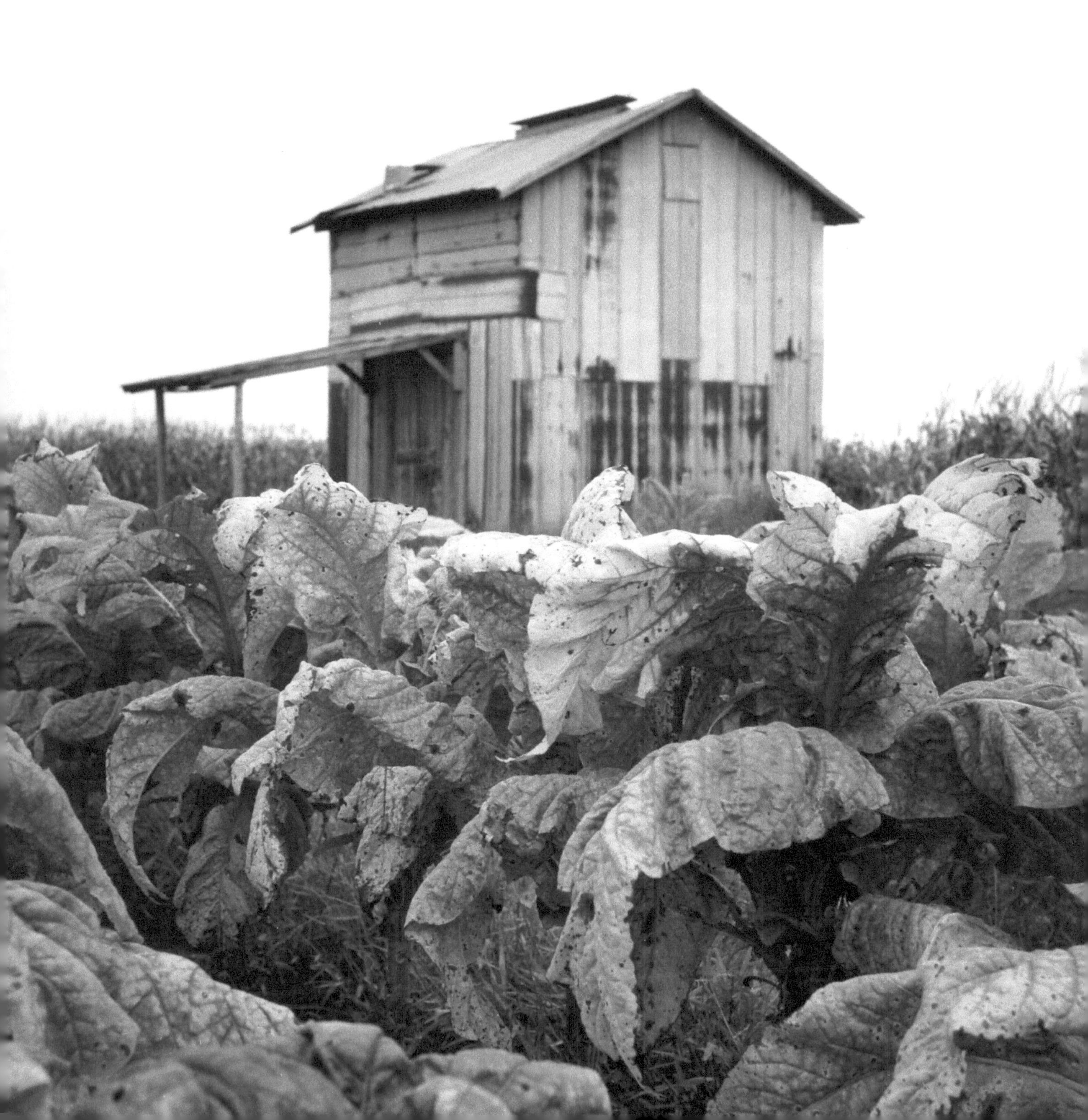

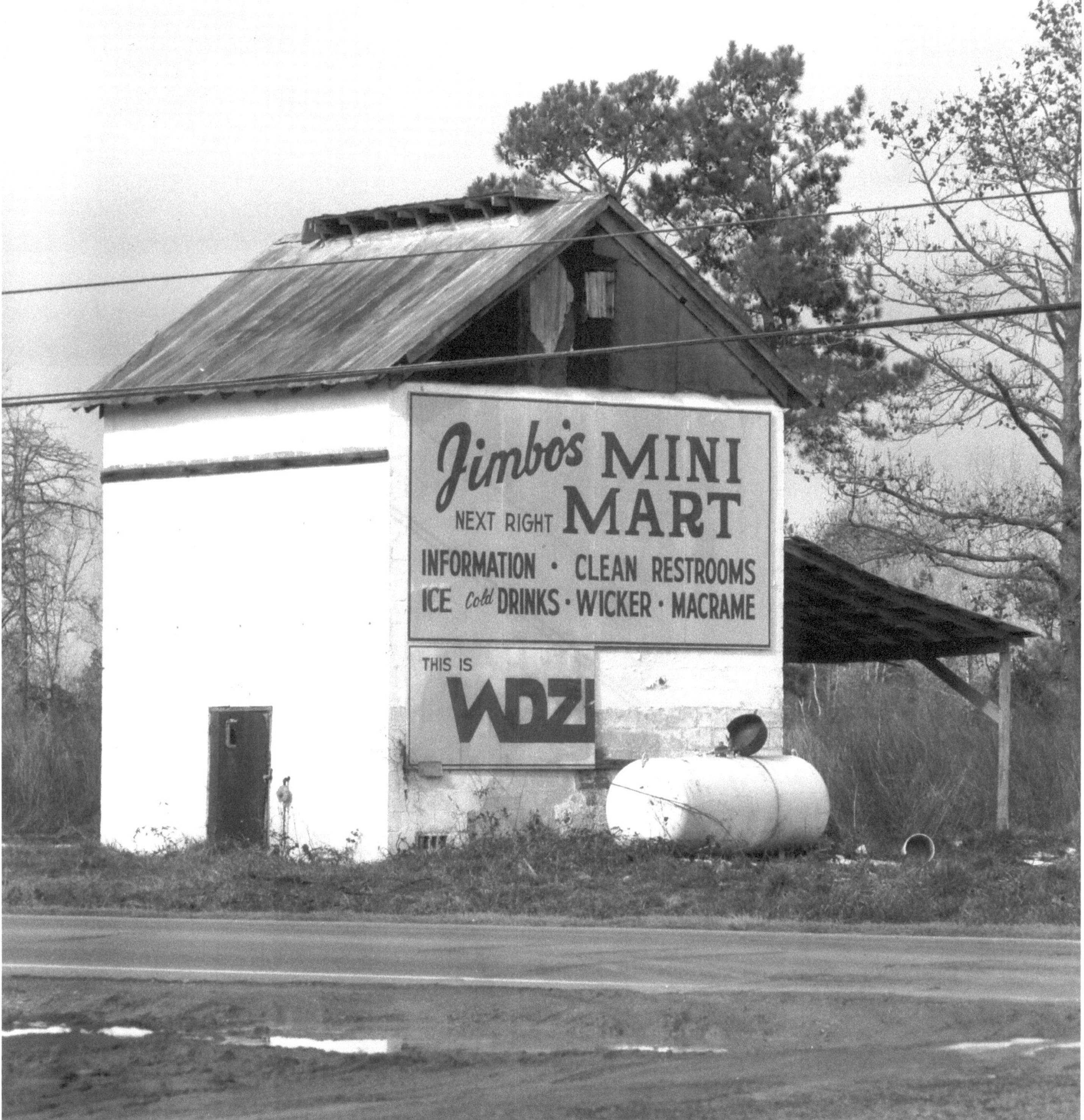
Jimbo's MINI MART
NEXT RIGHT
INFORMATION • CLEAN RESTROOMS
ICE Cold DRINKS • WICKER • MACRAME
THIS IS
WDZ

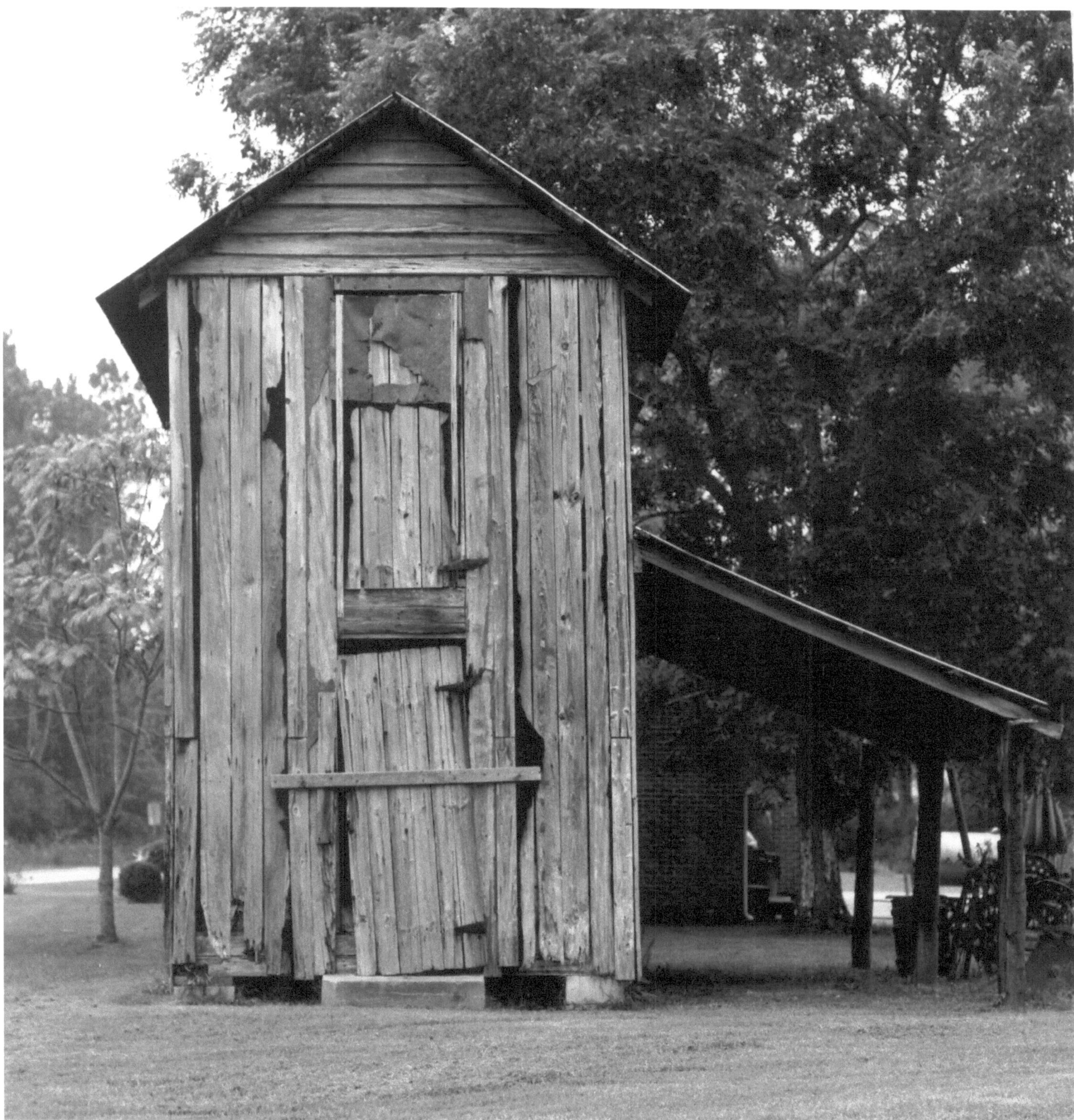

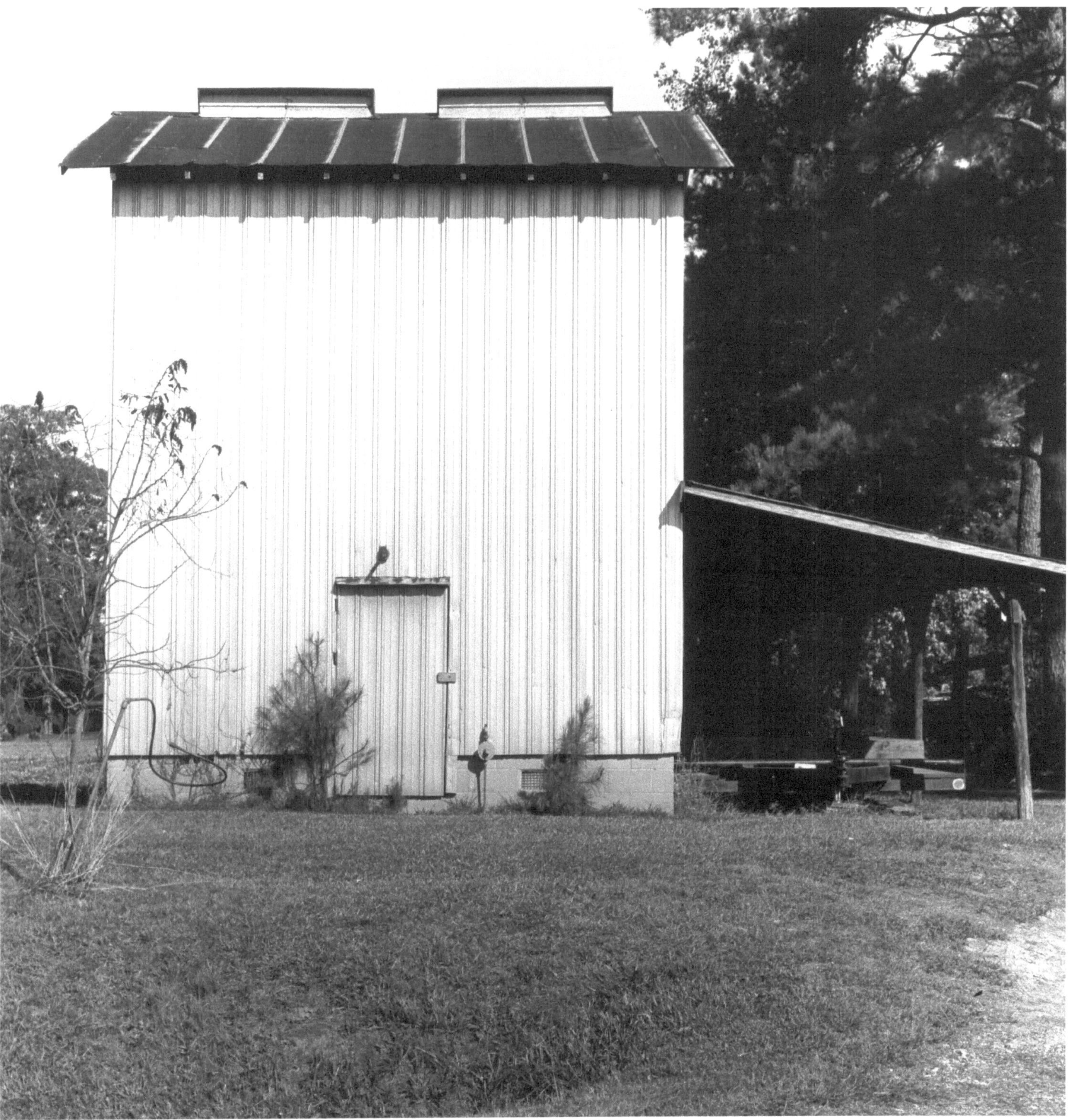

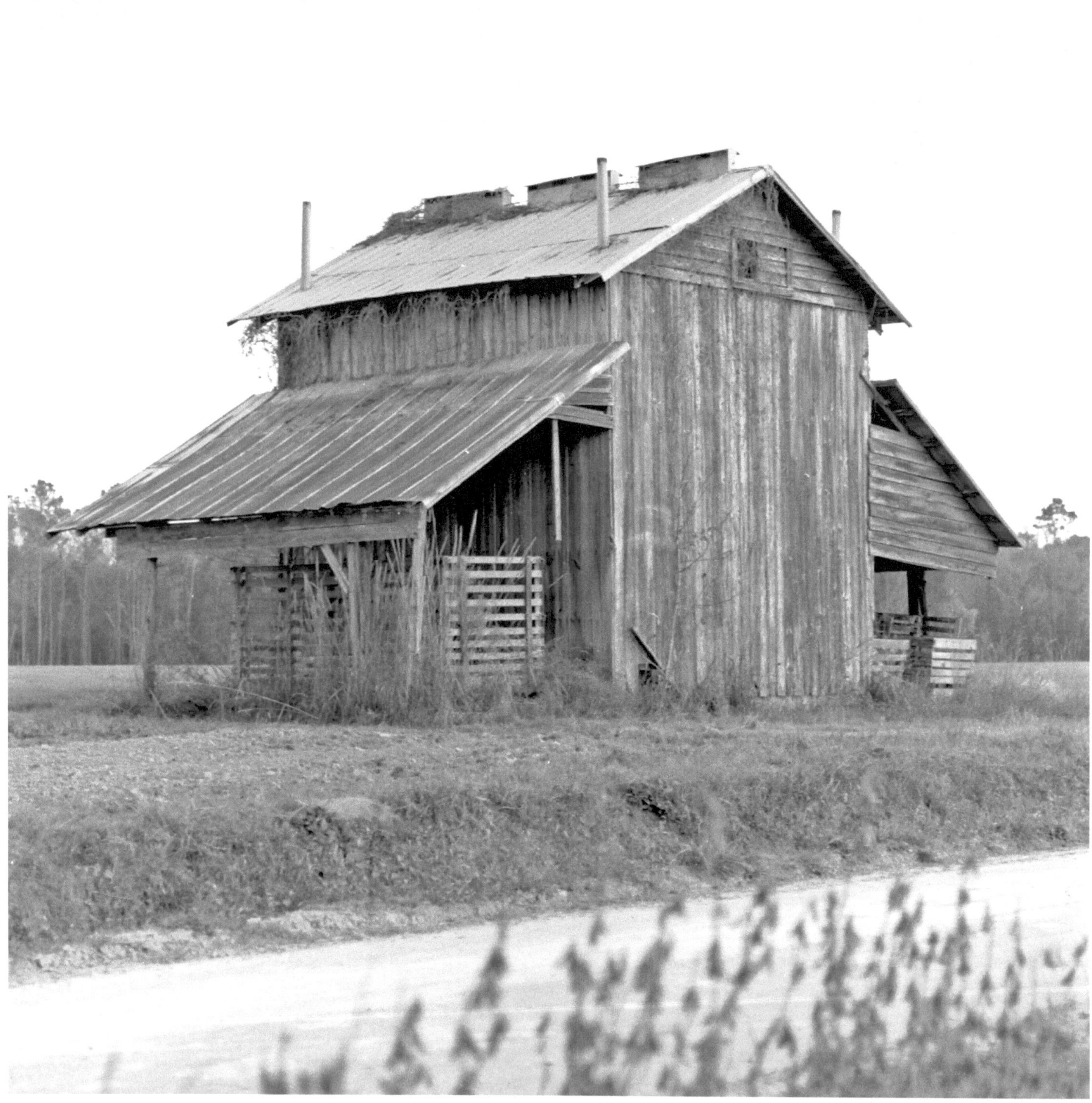

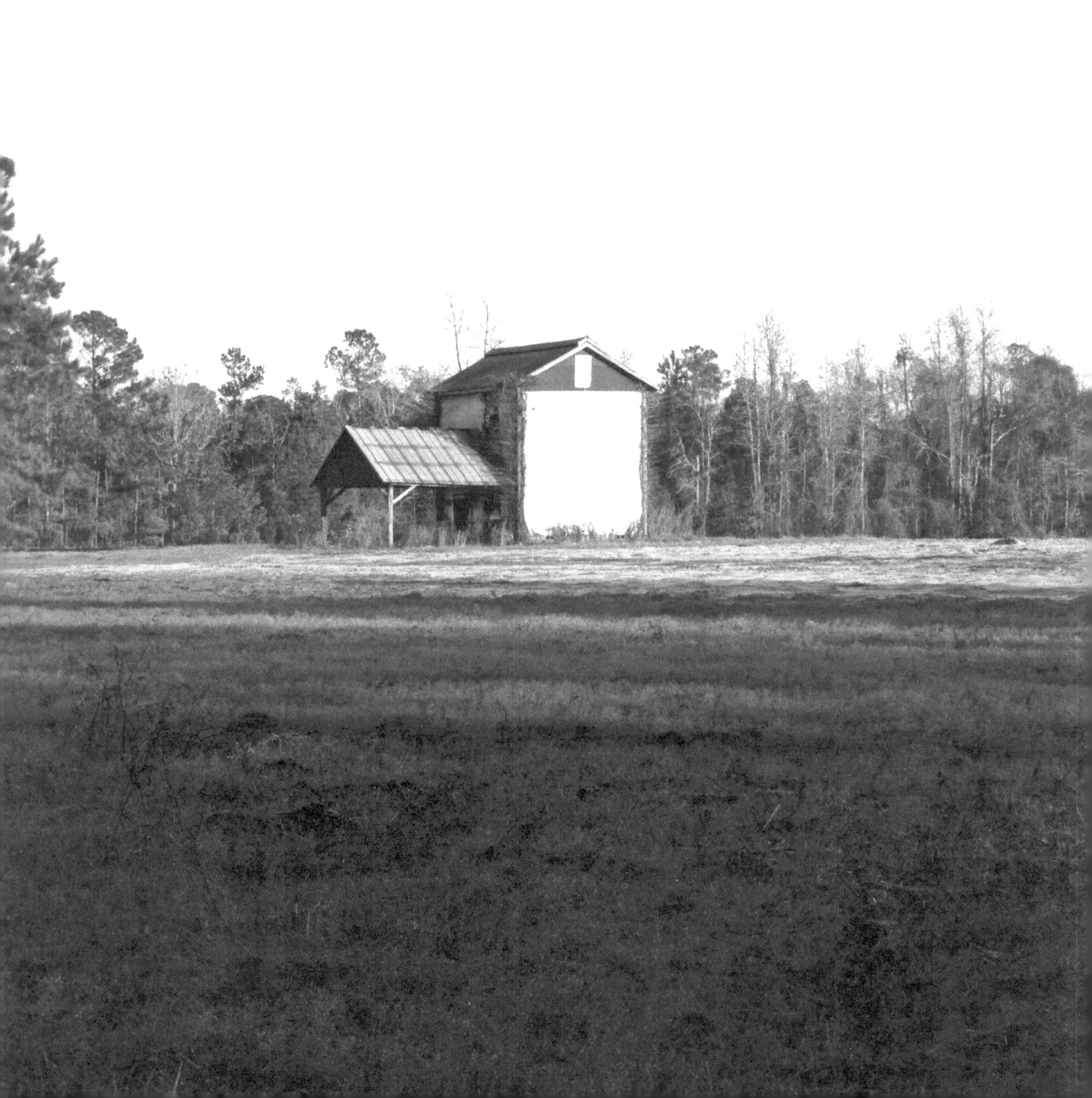

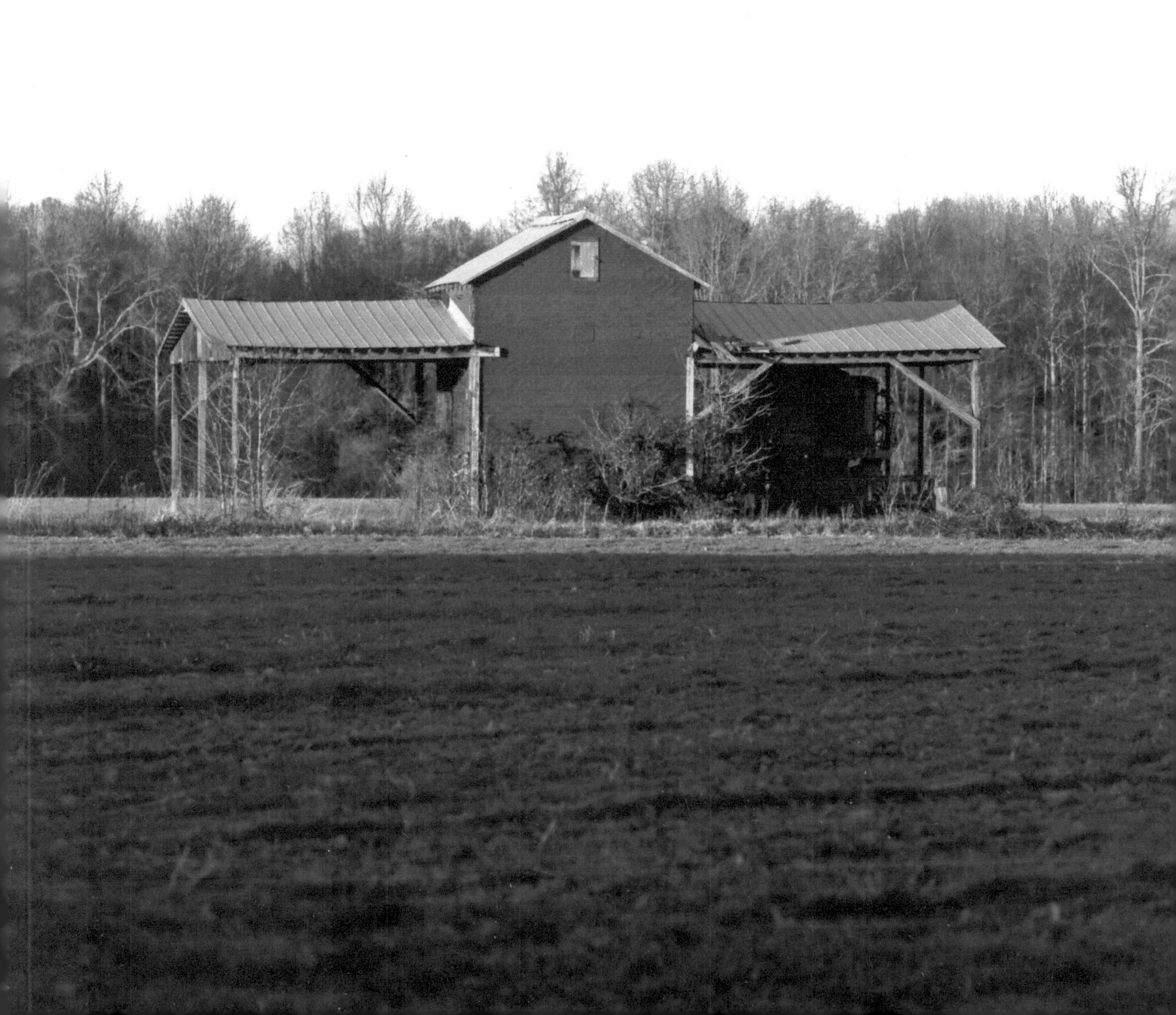

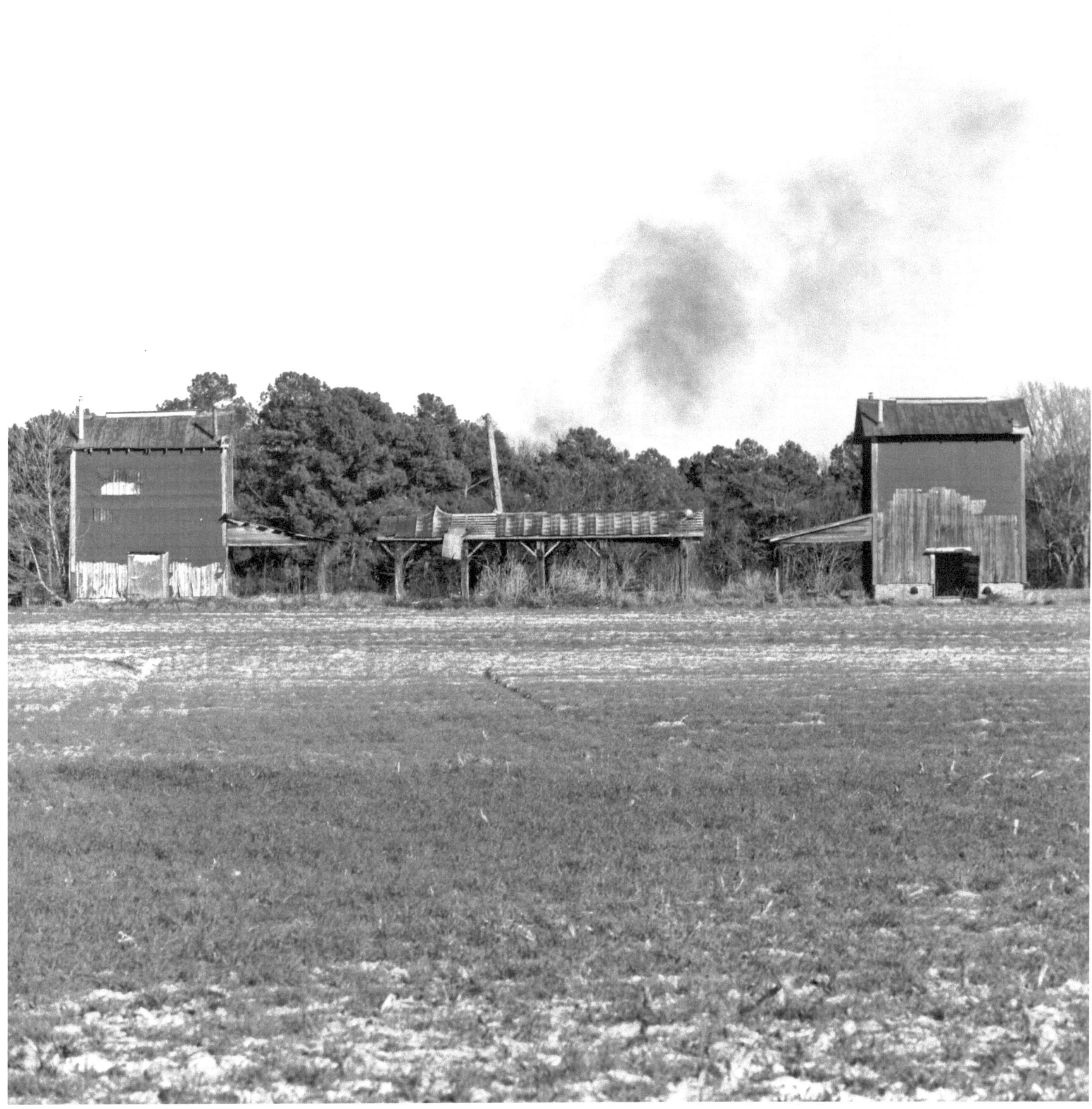

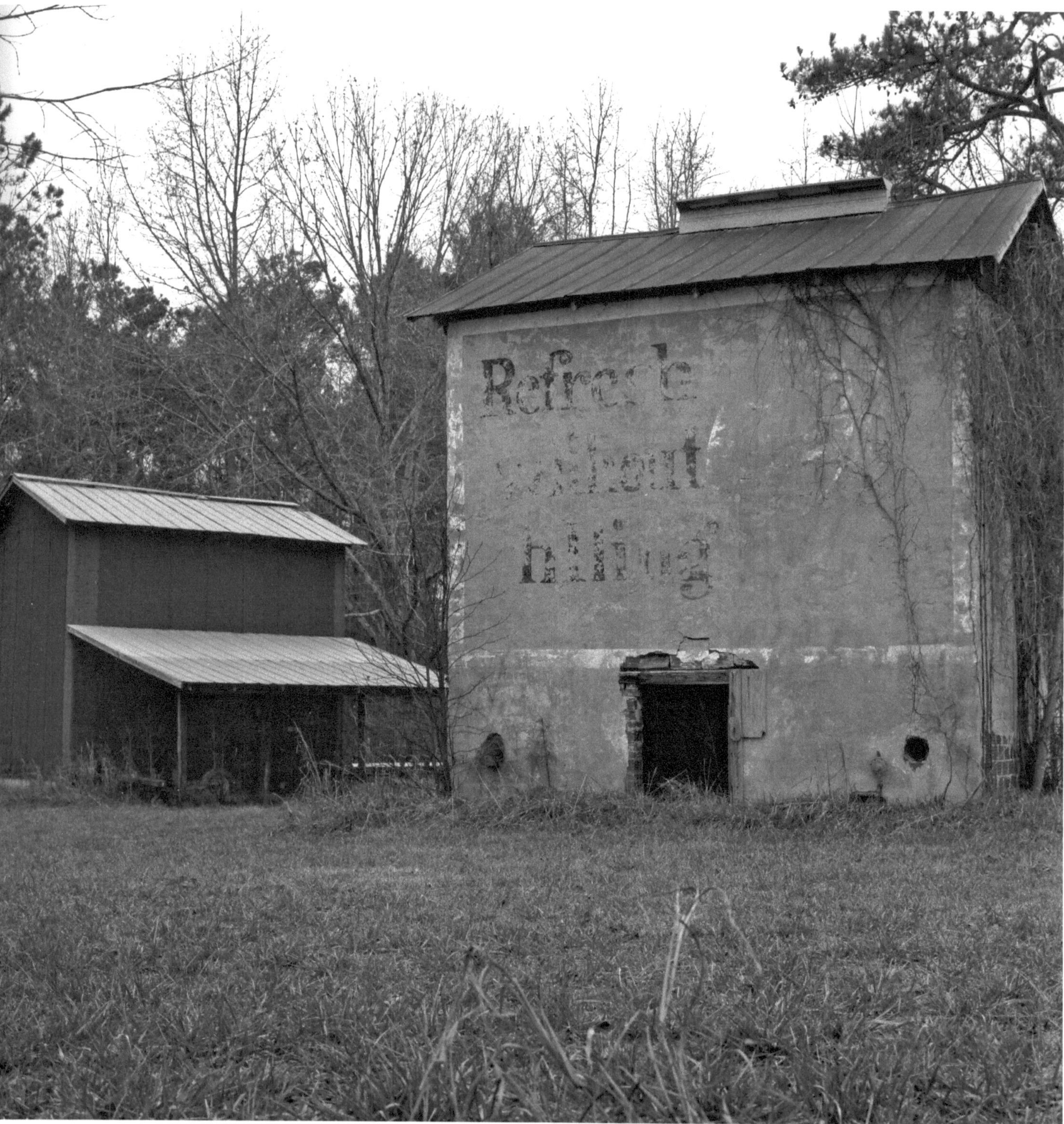

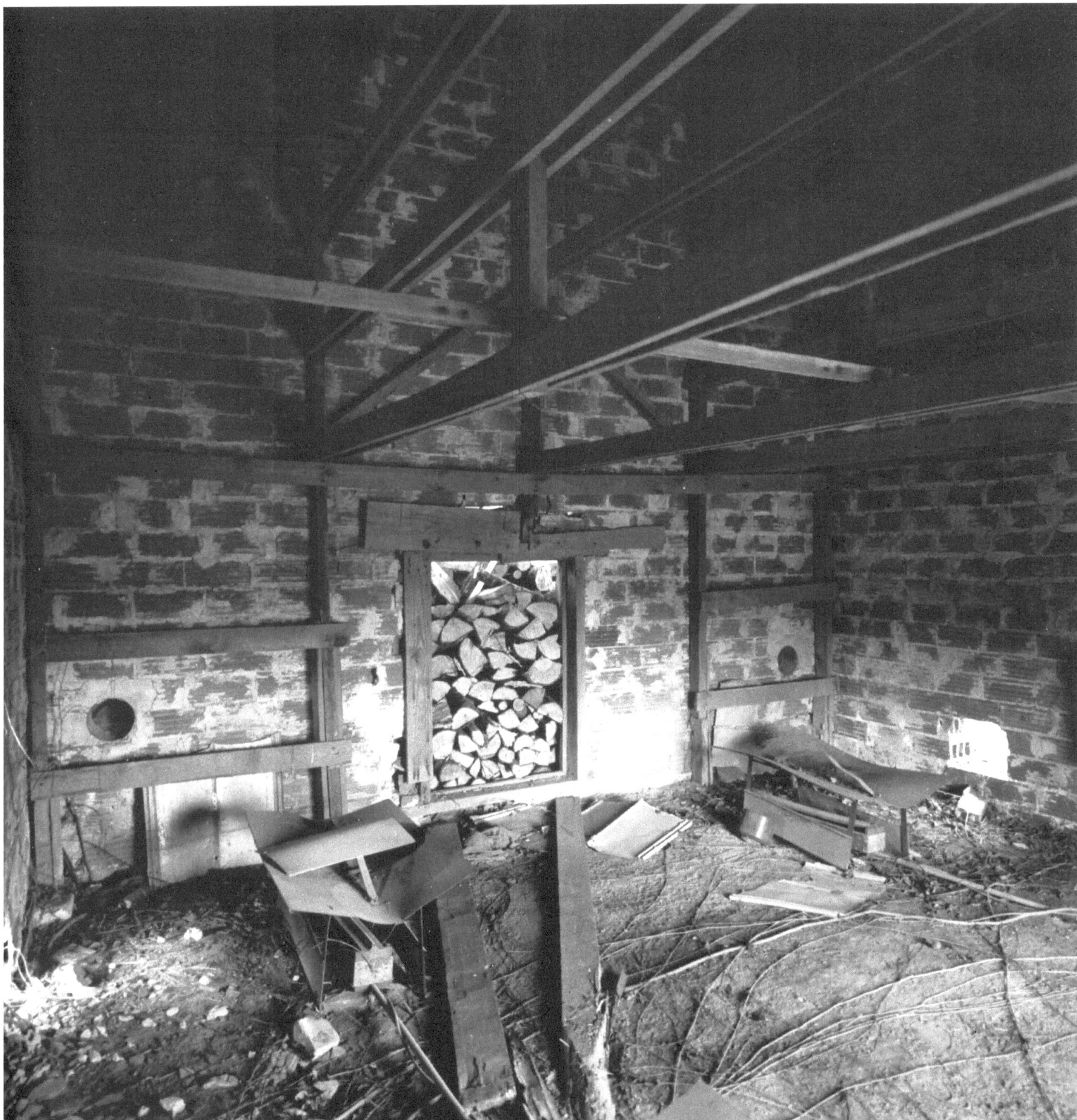

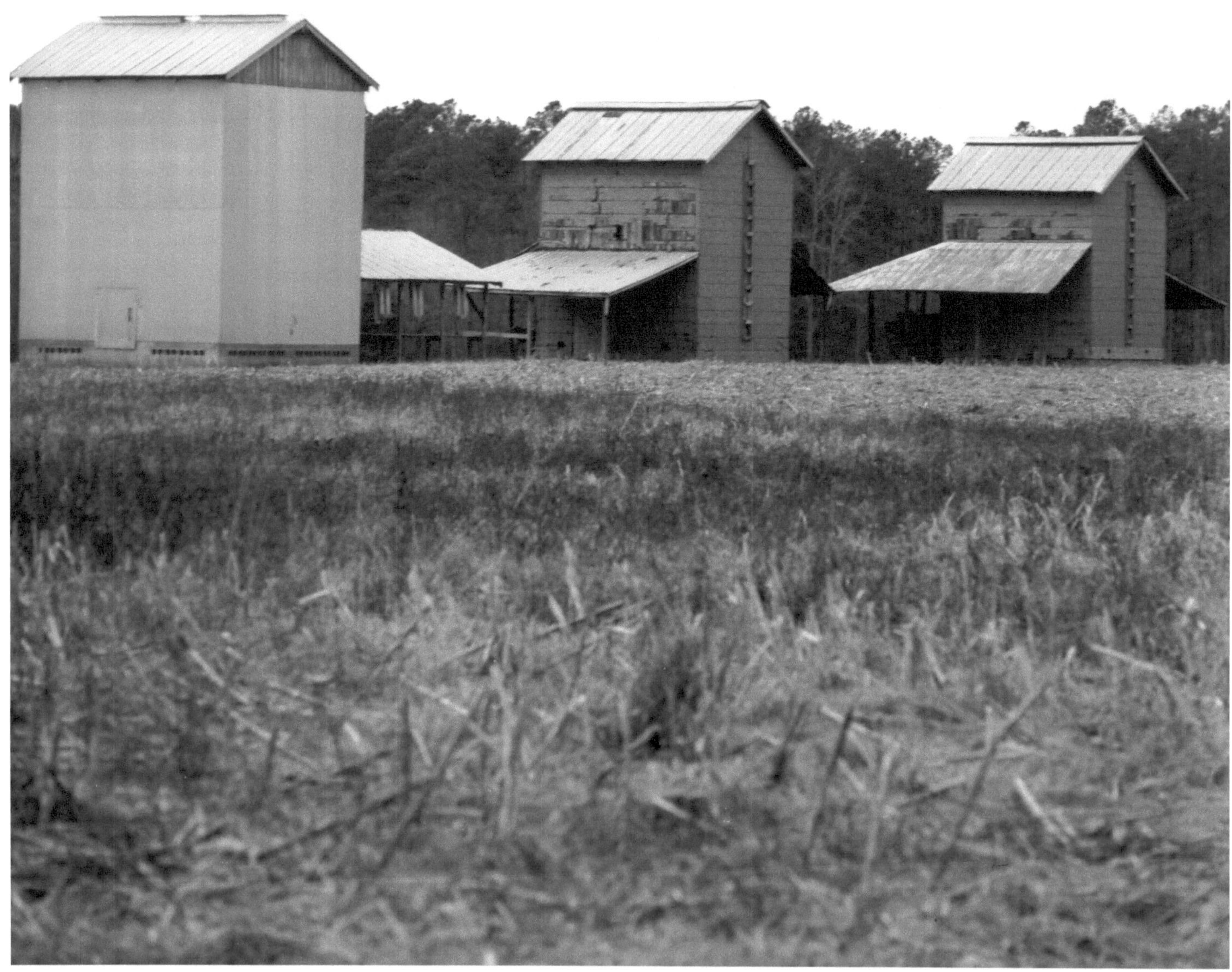

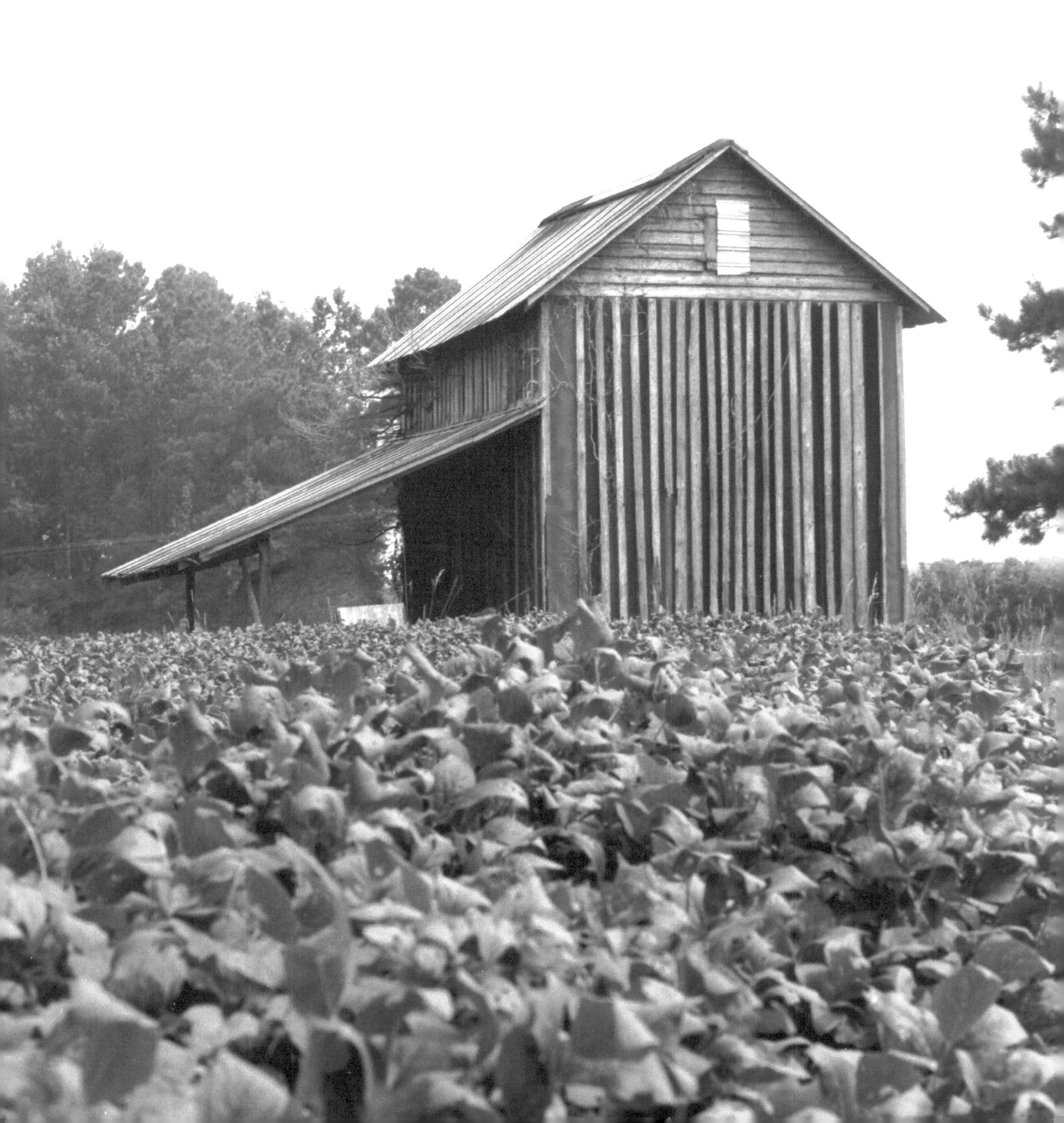

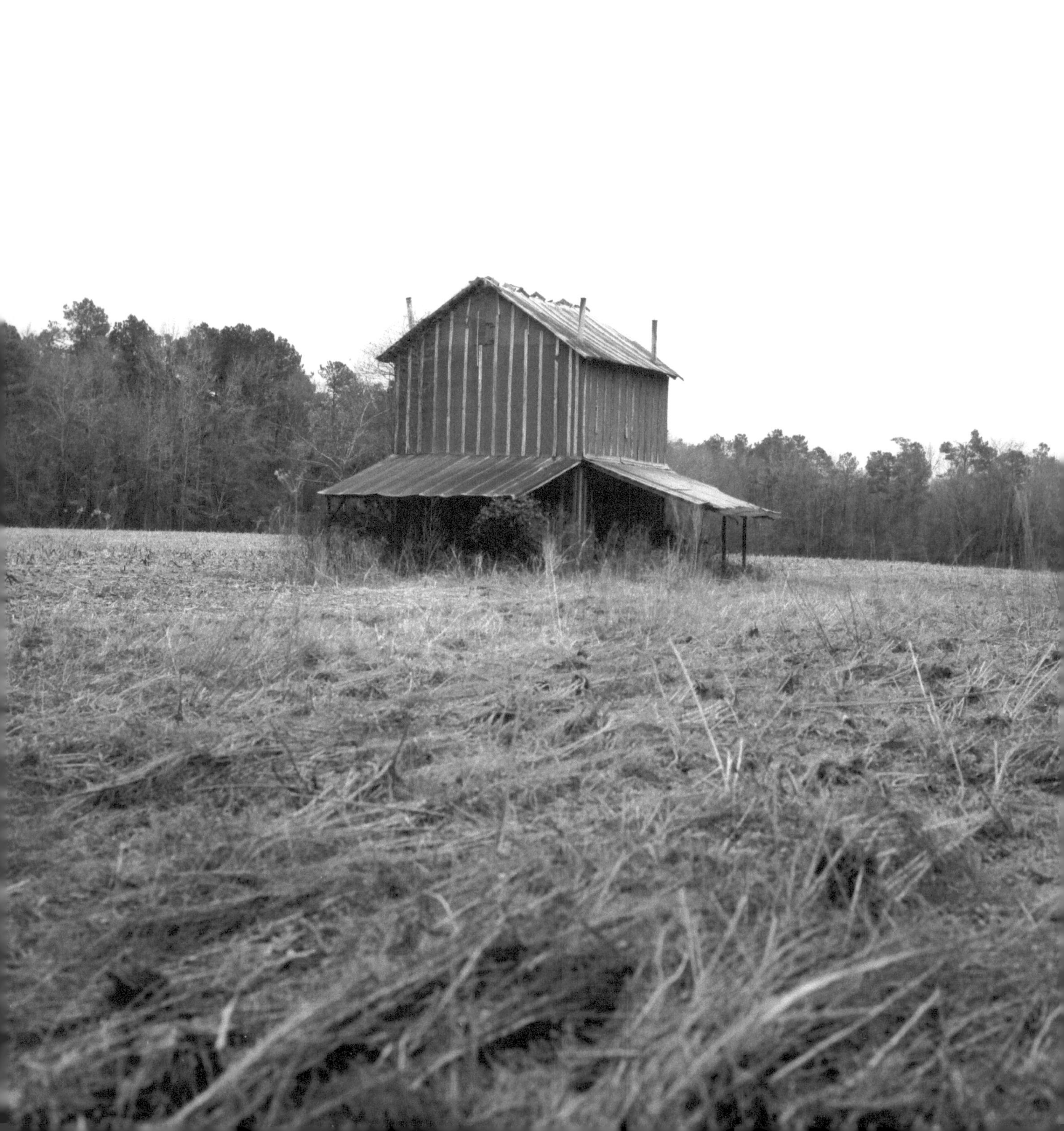

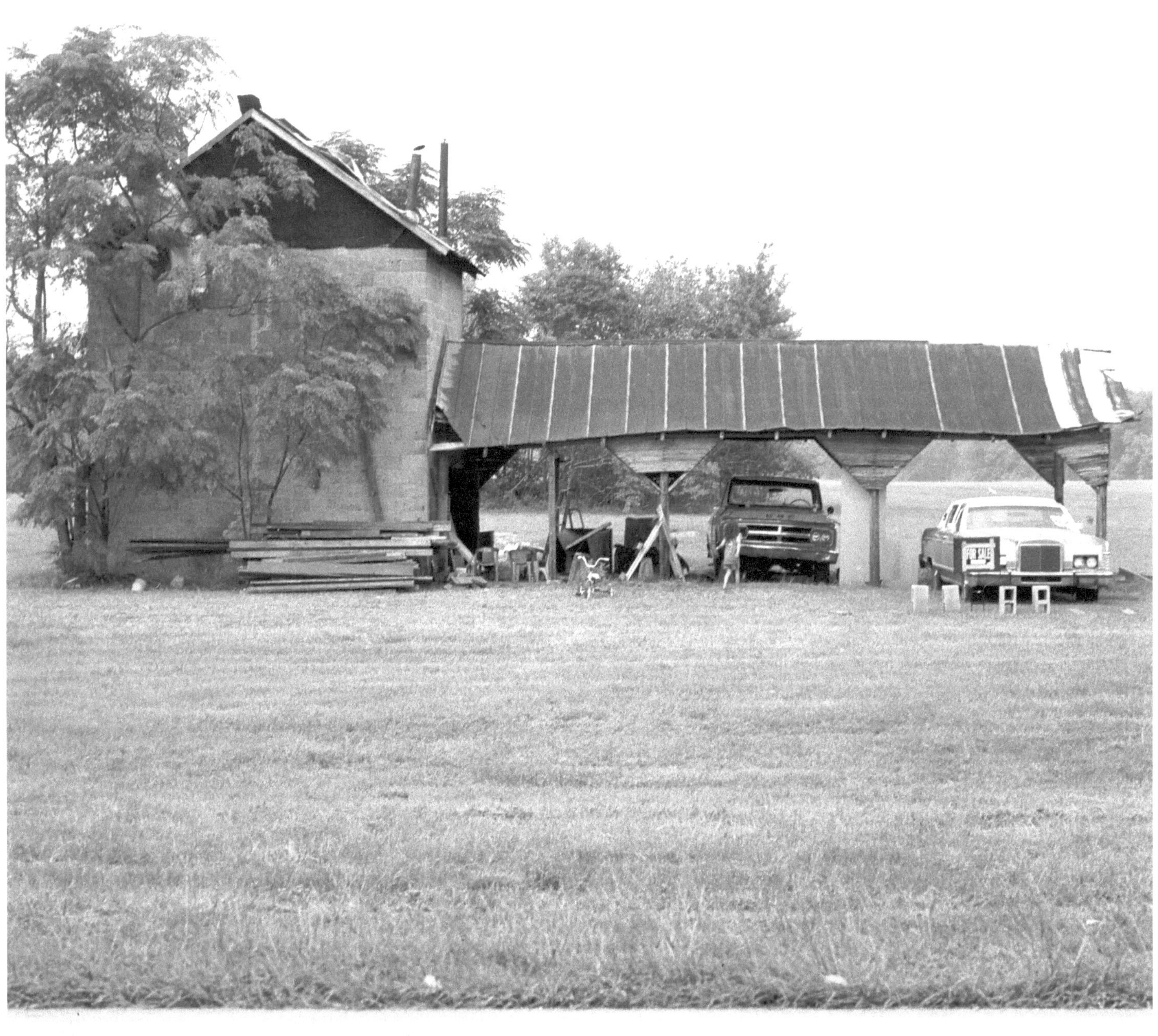
FOR SALE

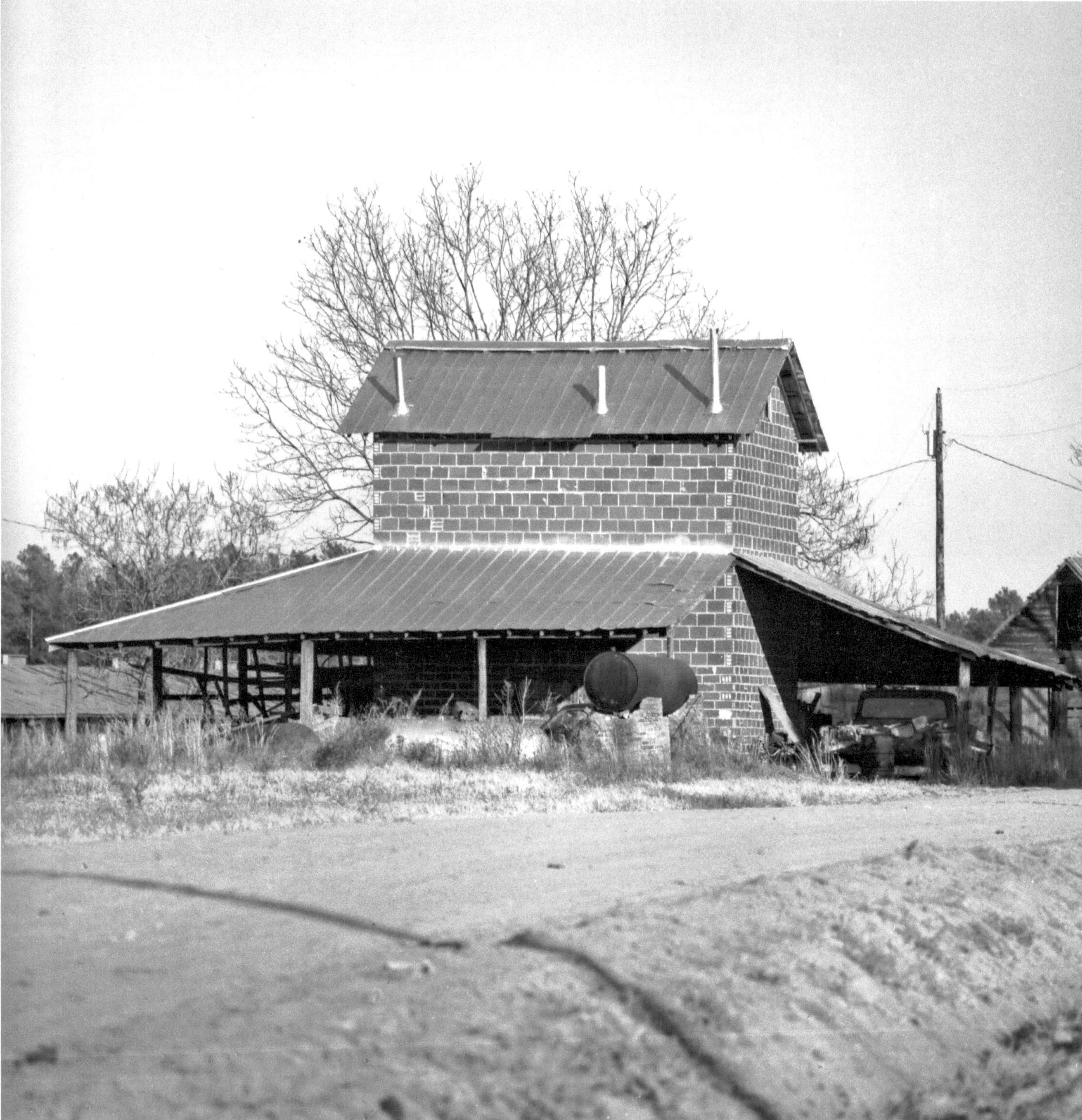

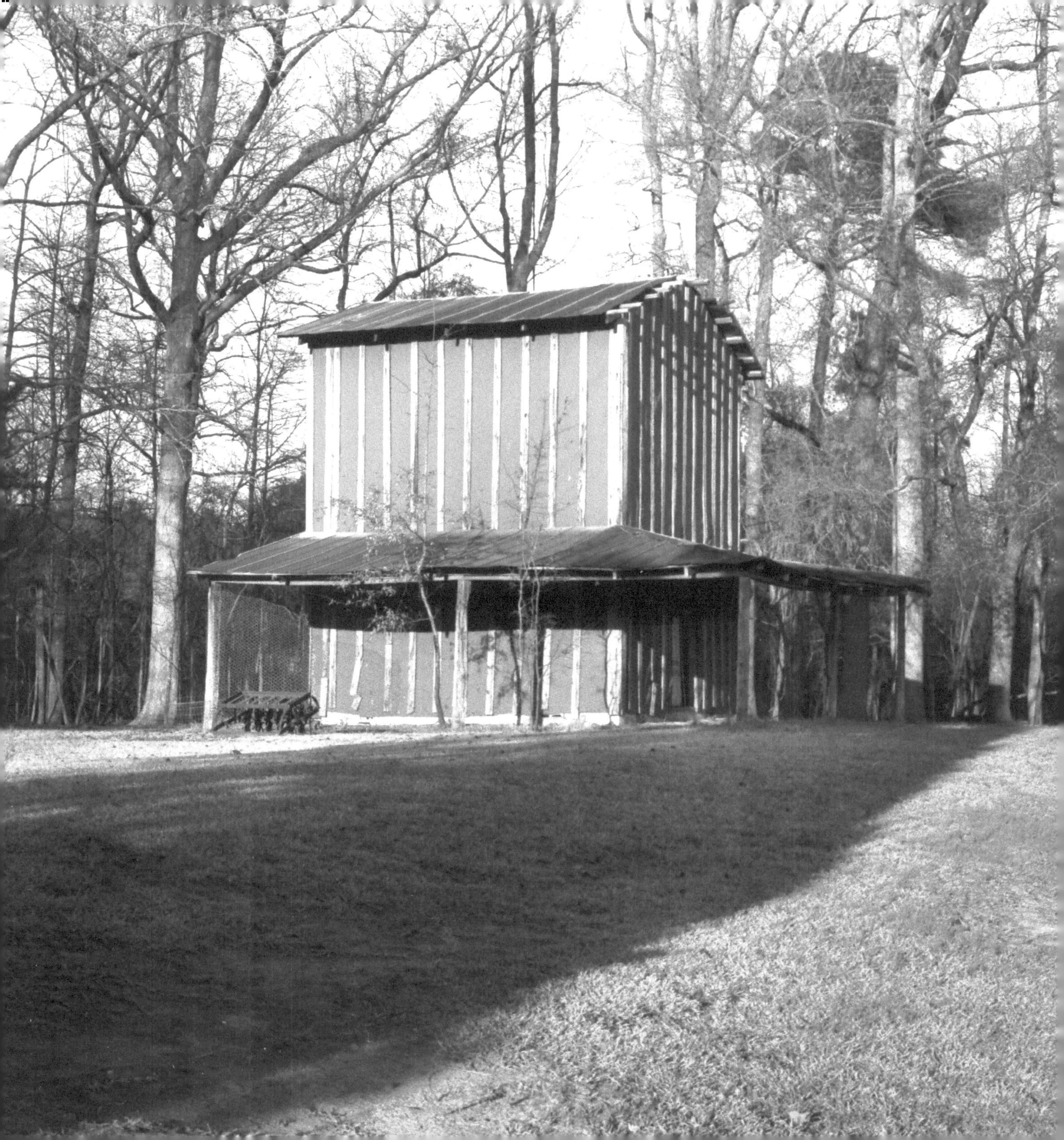

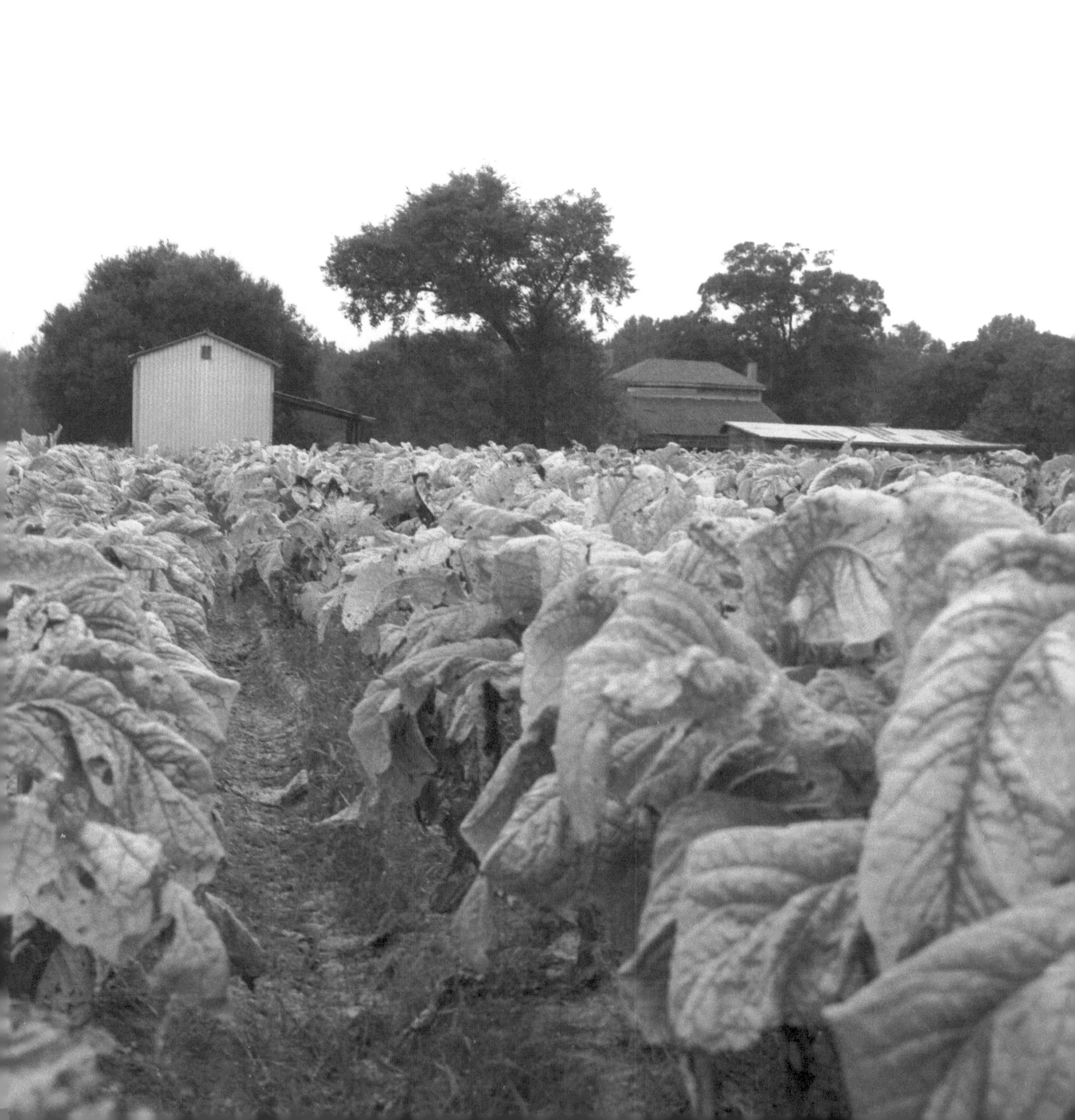

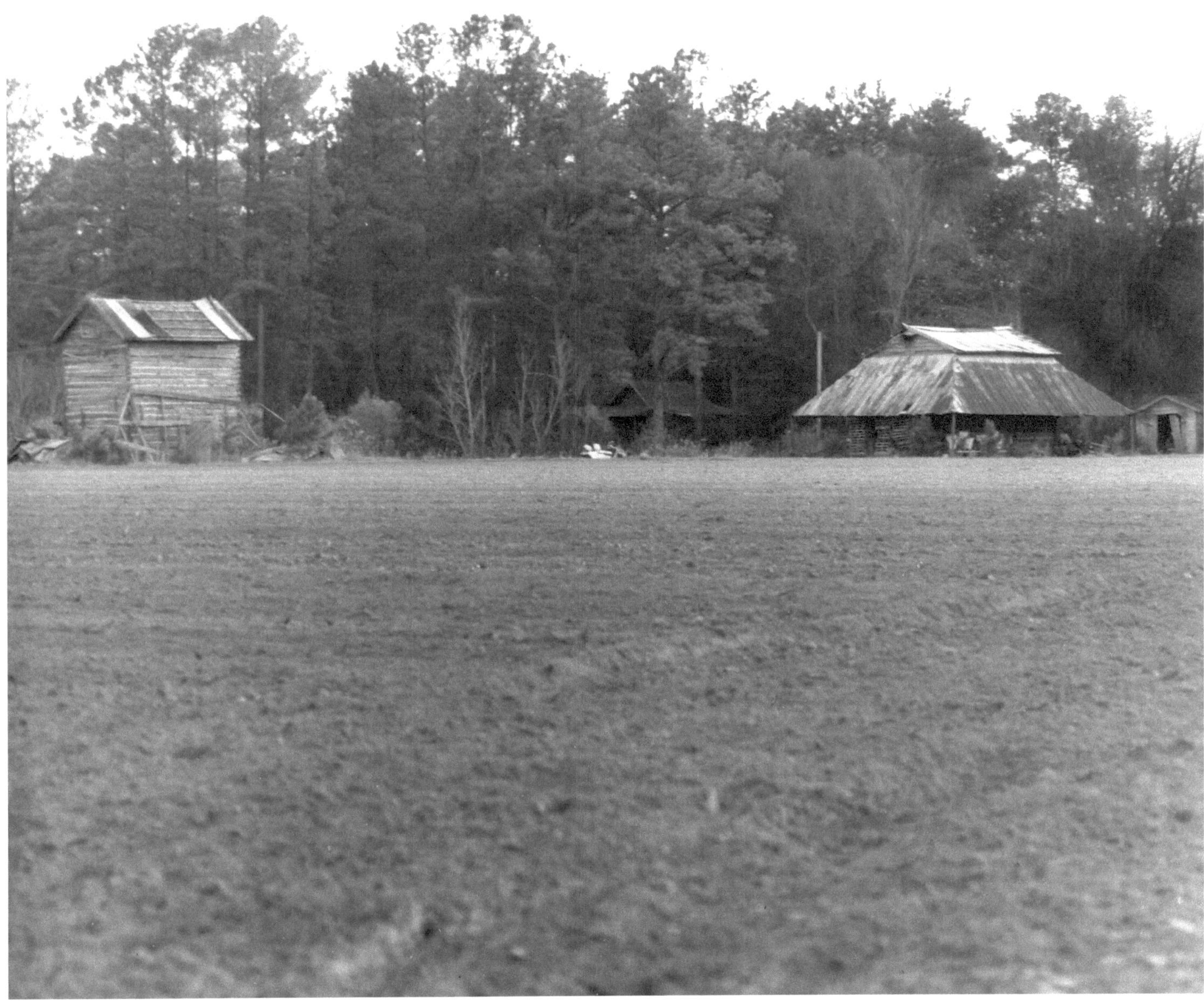

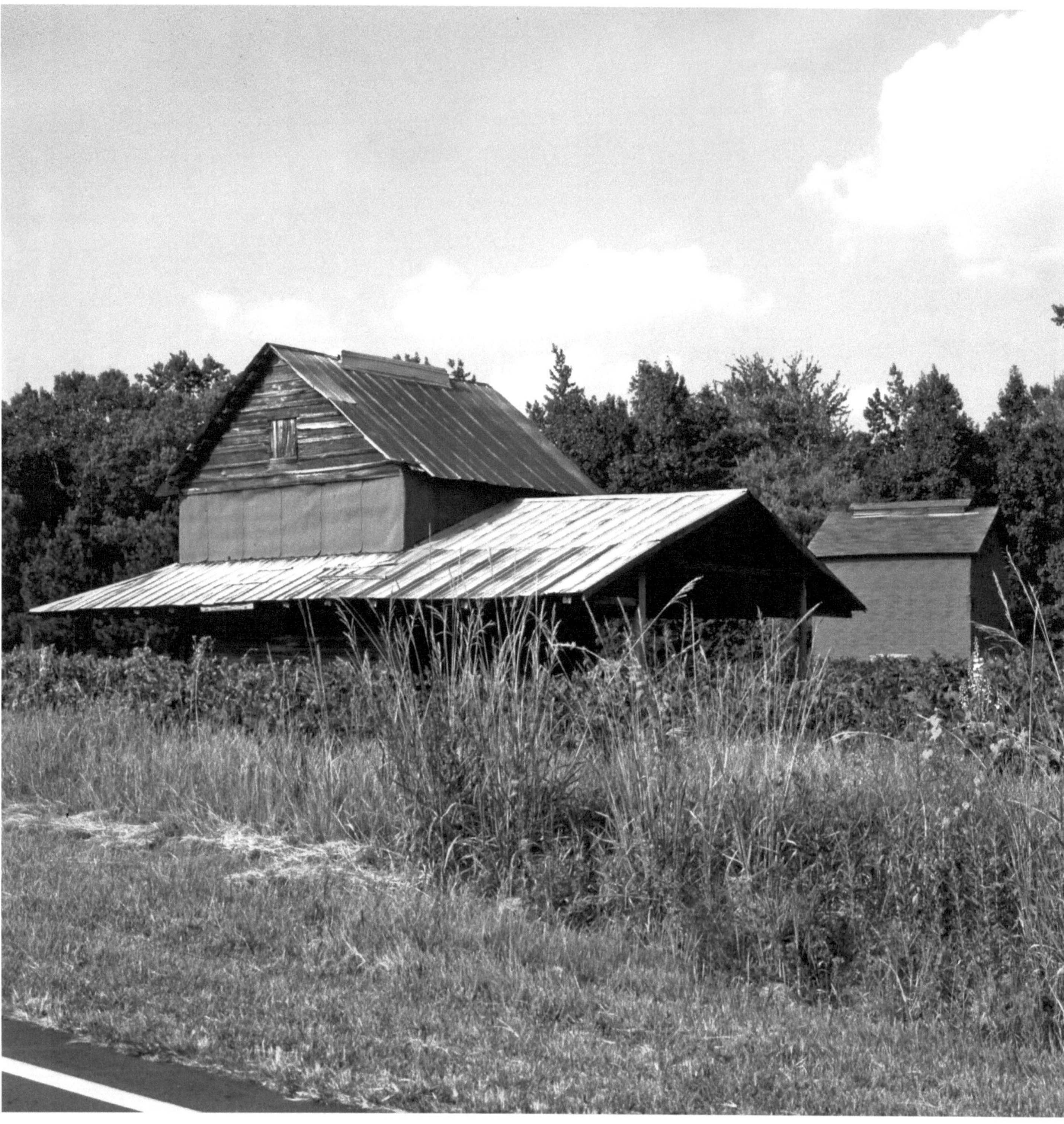

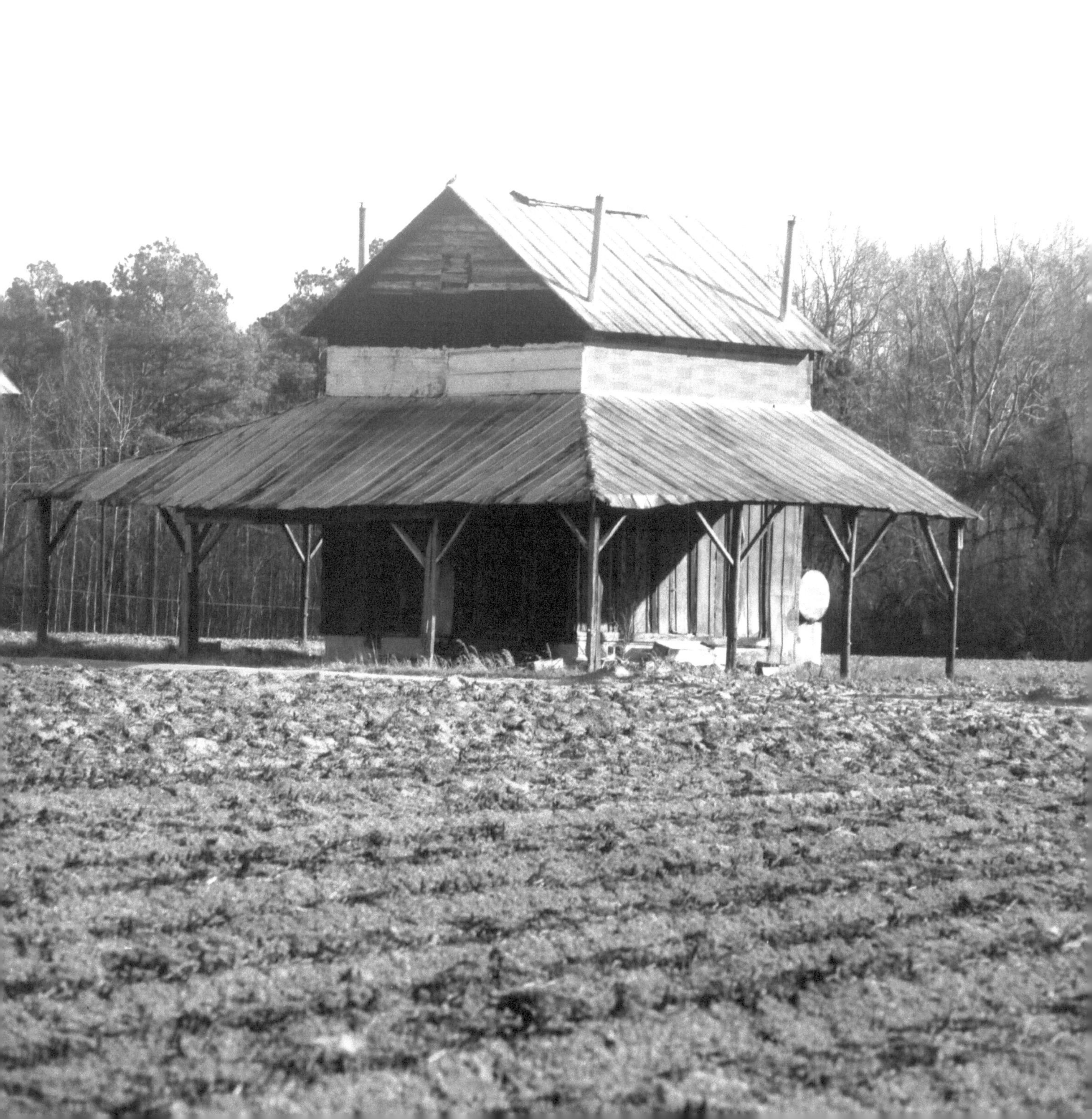

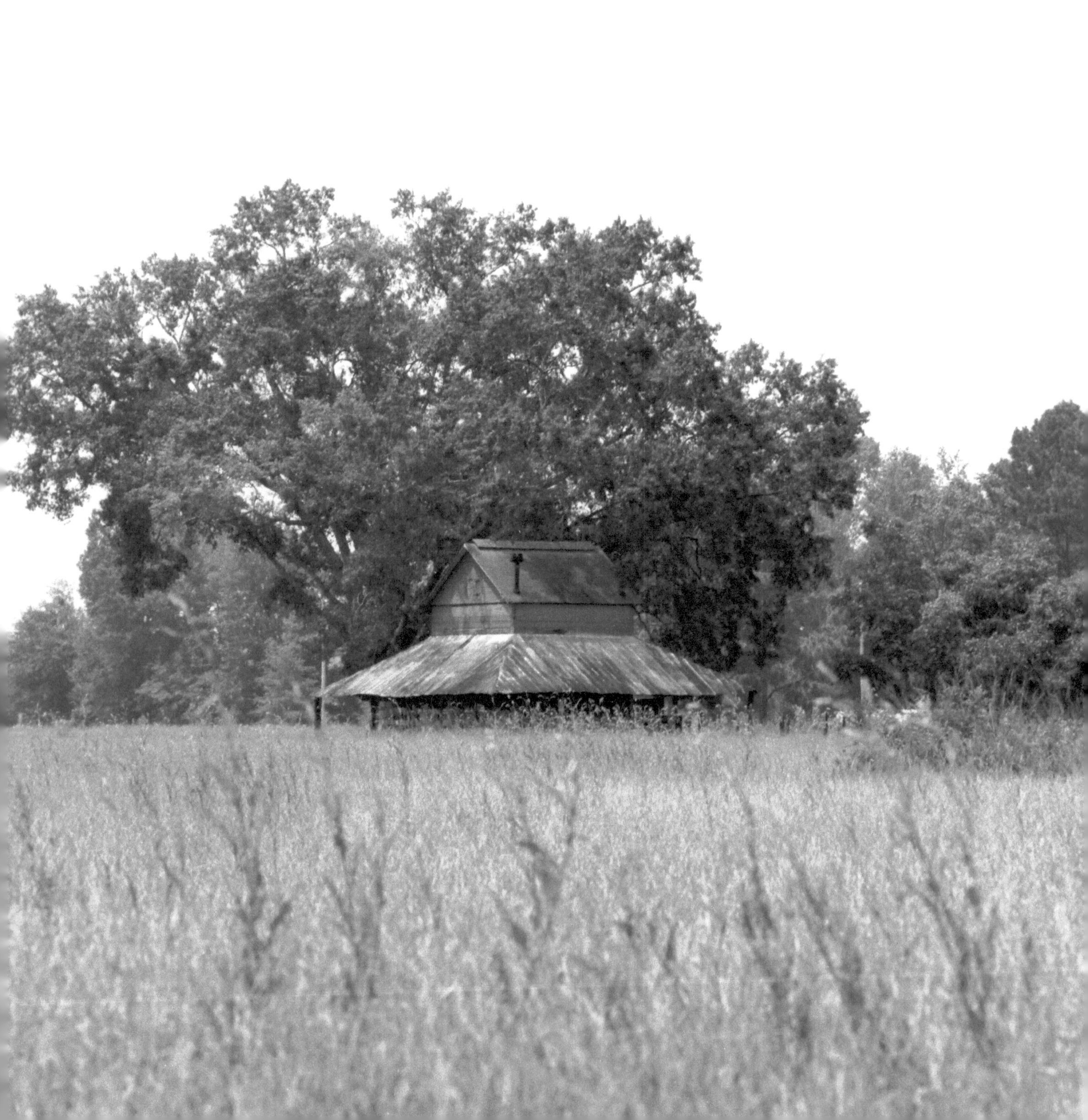

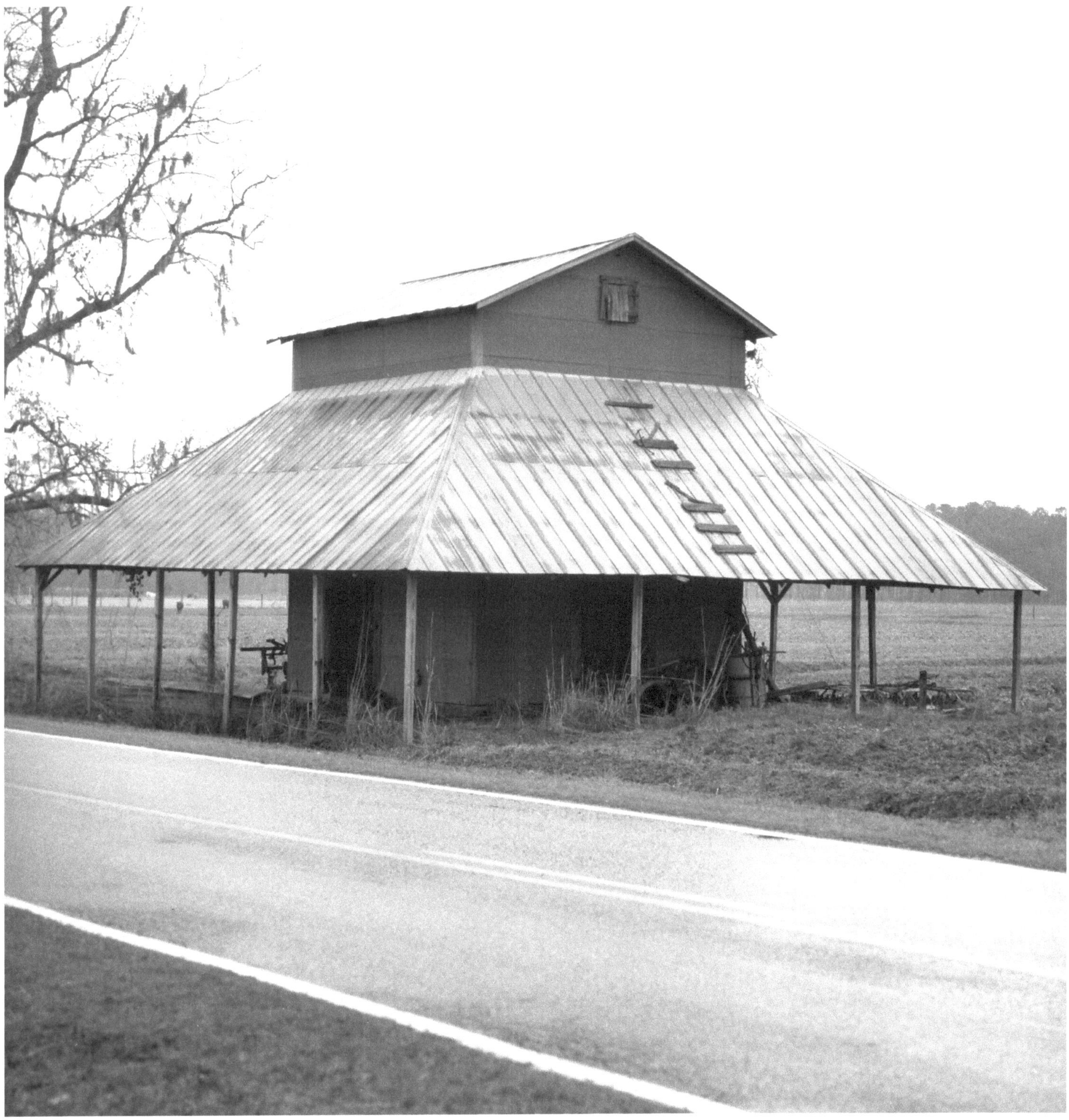

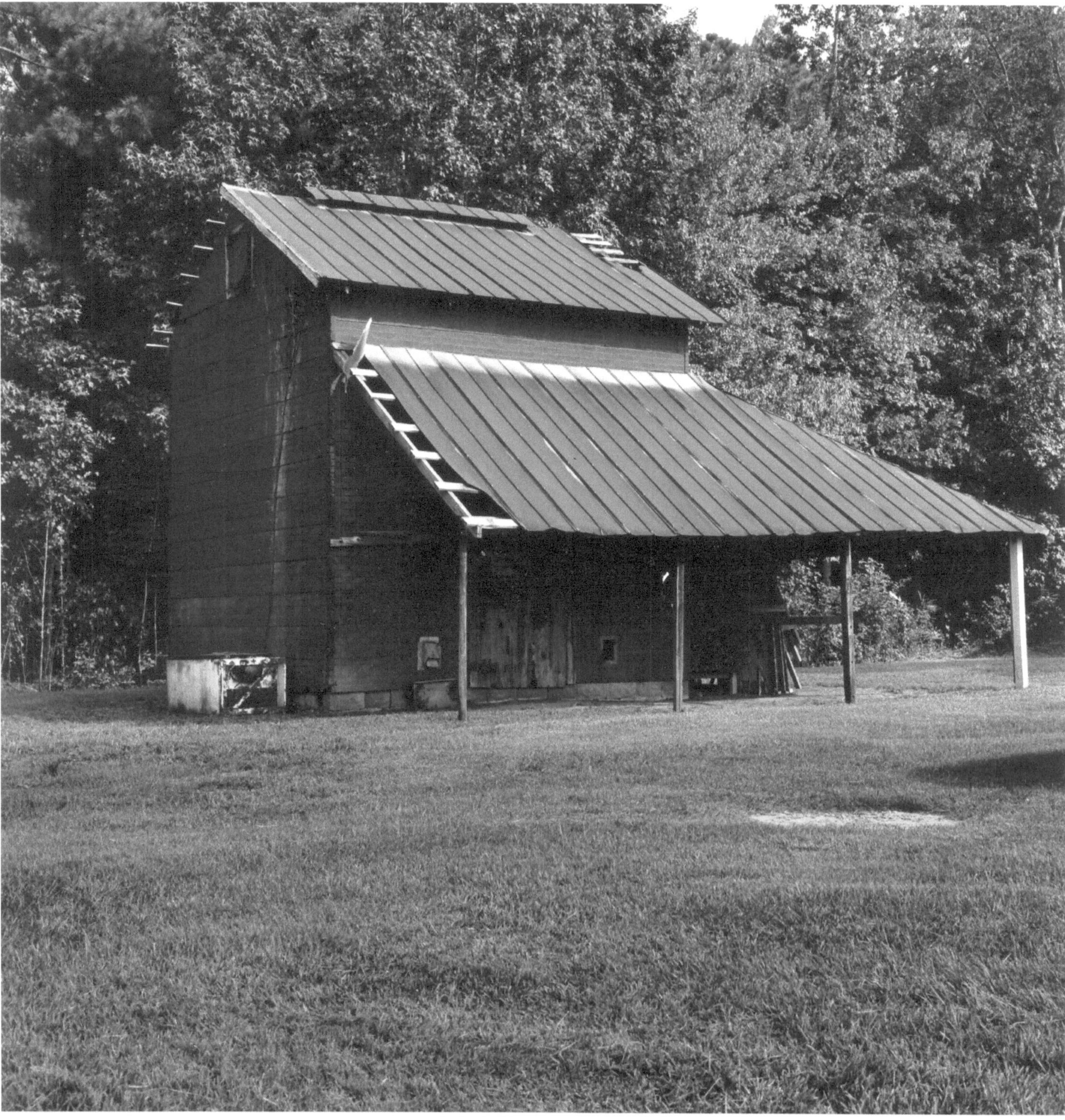

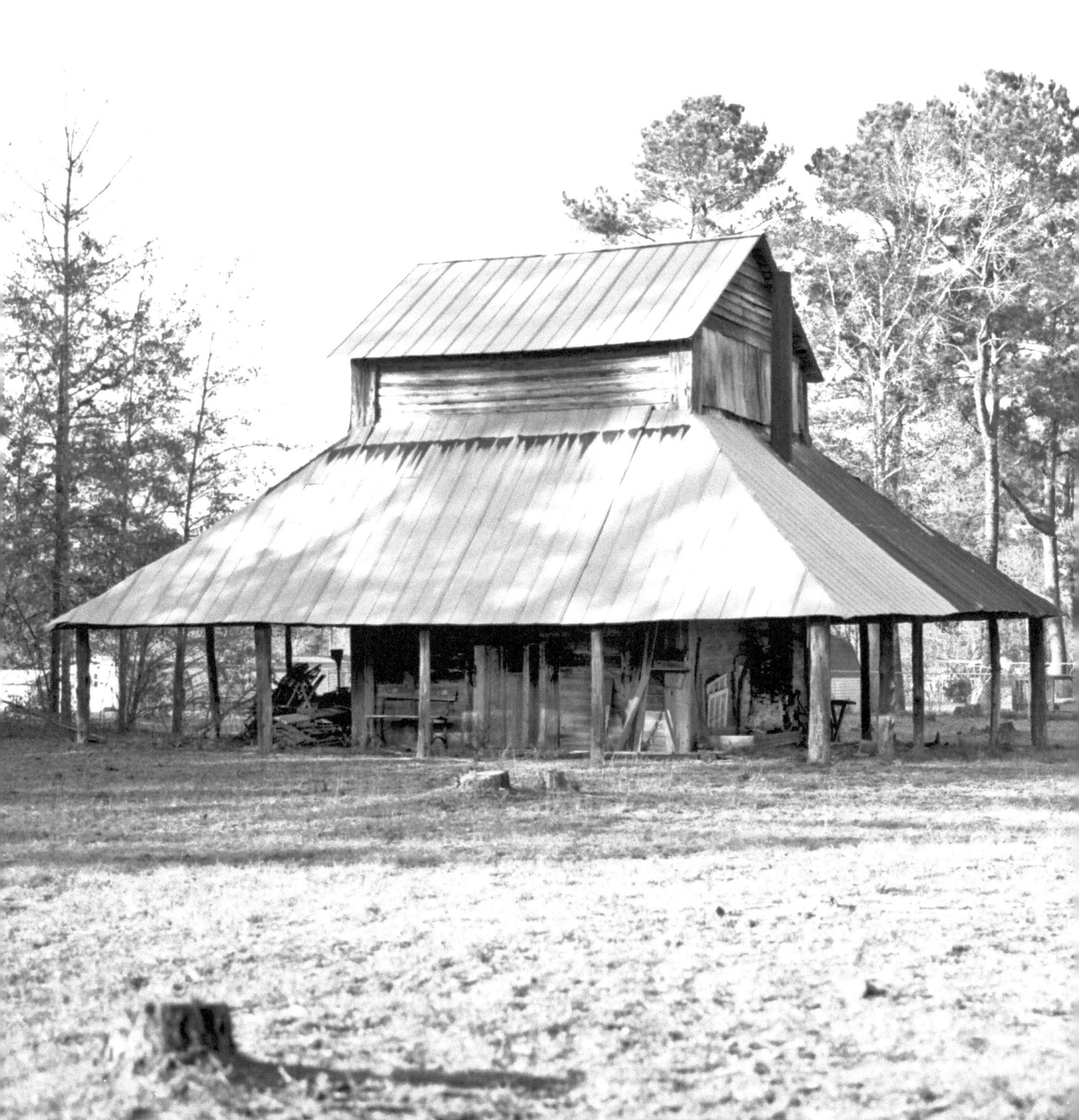

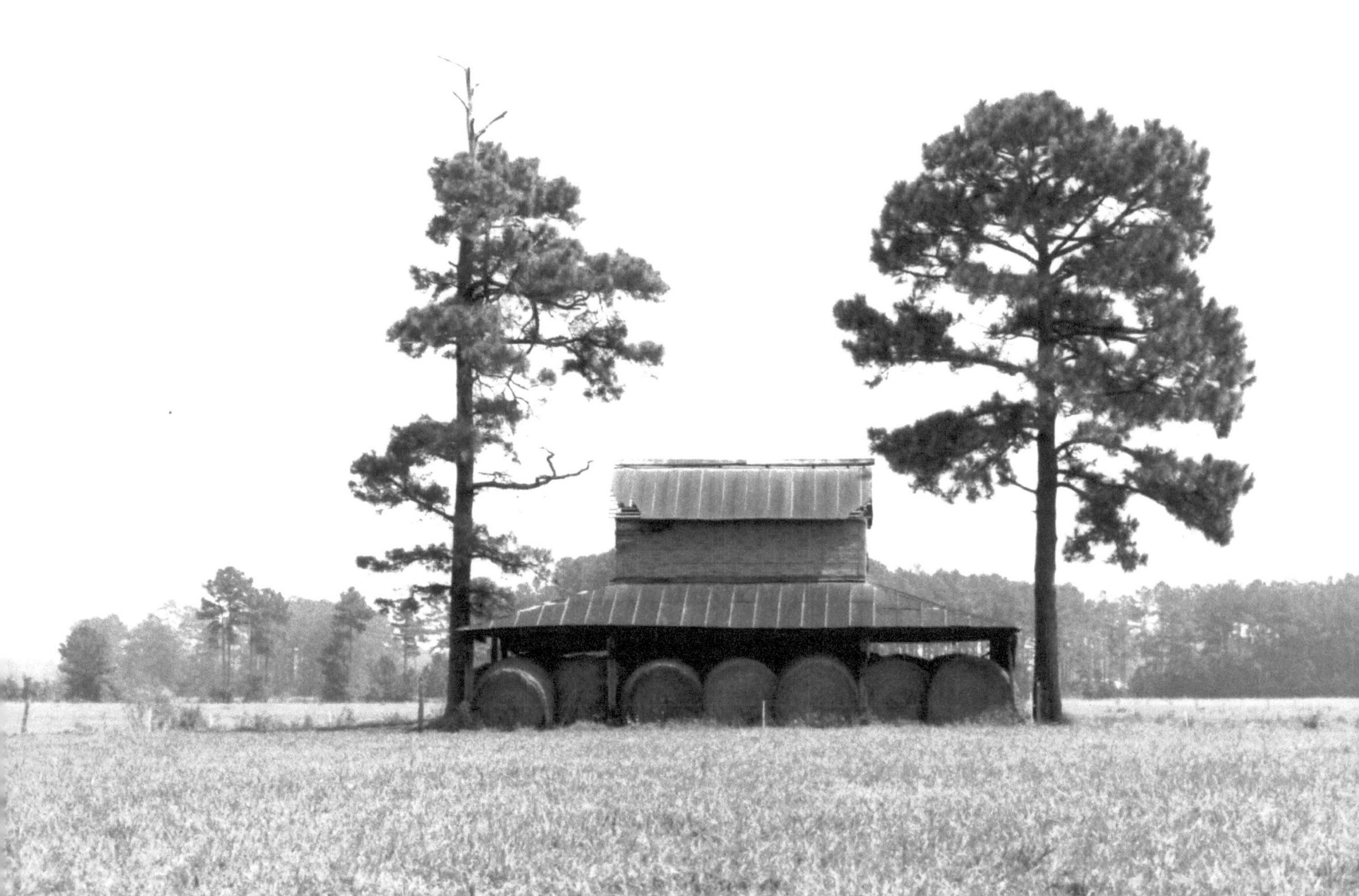

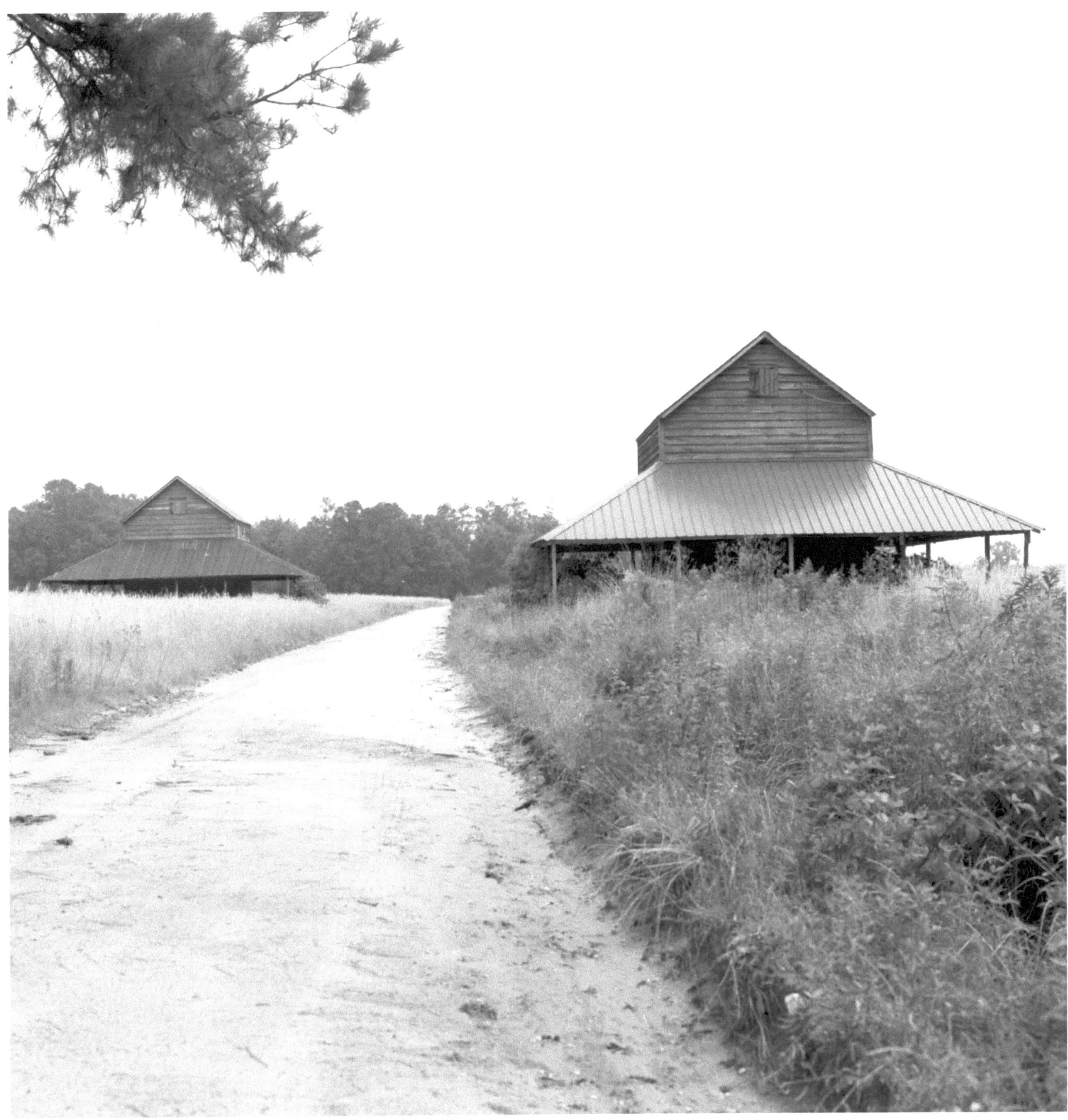

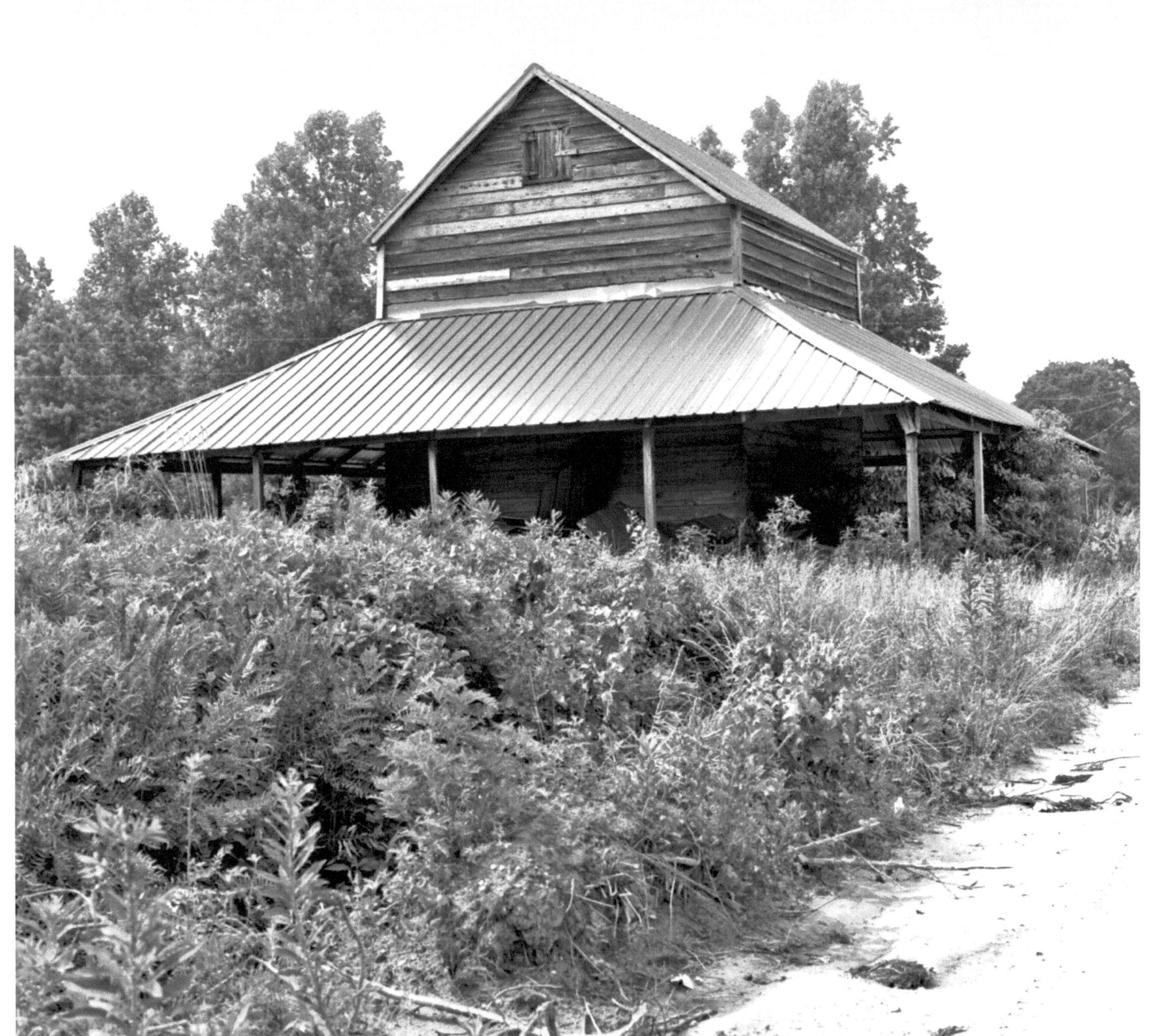

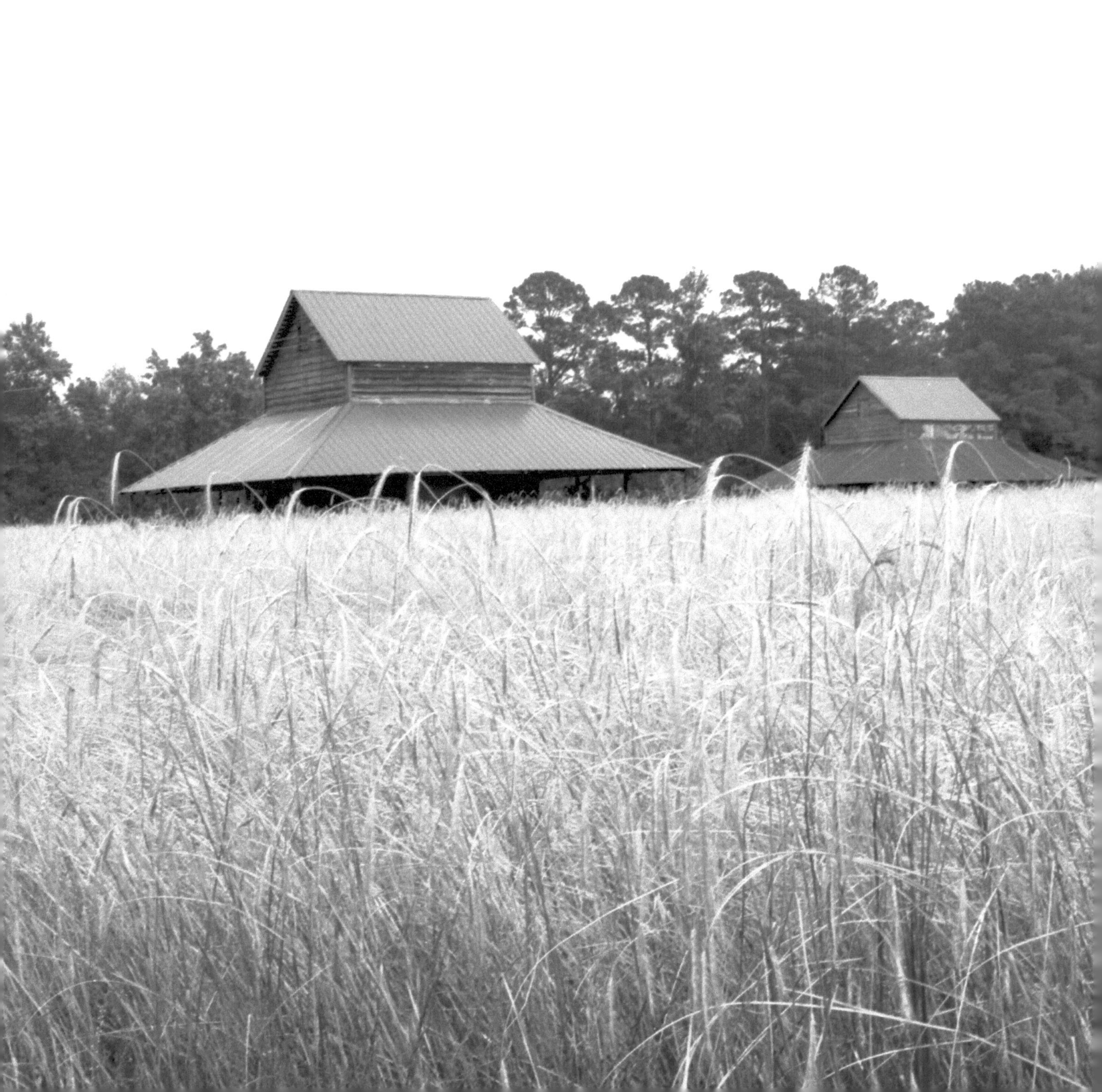

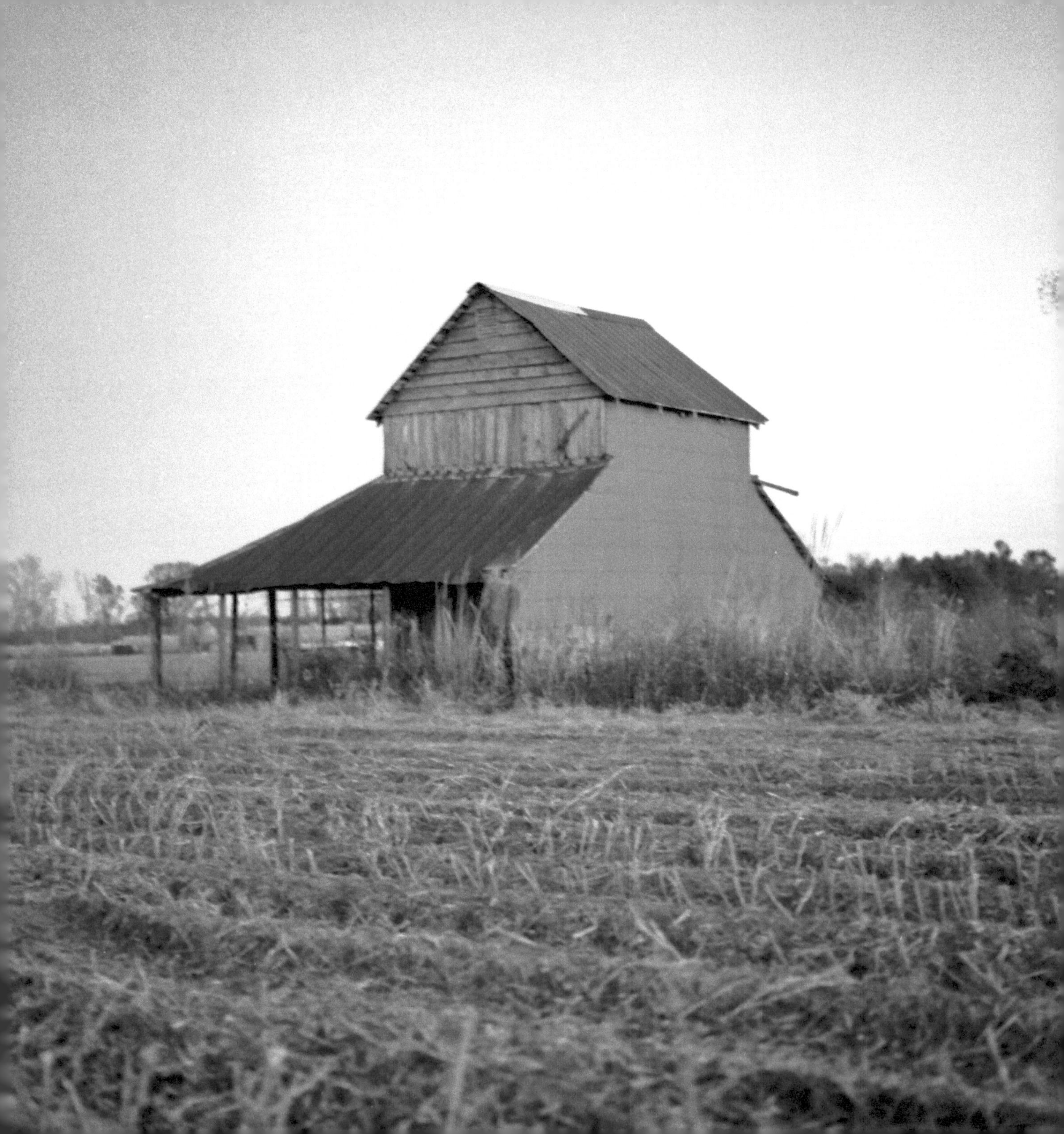

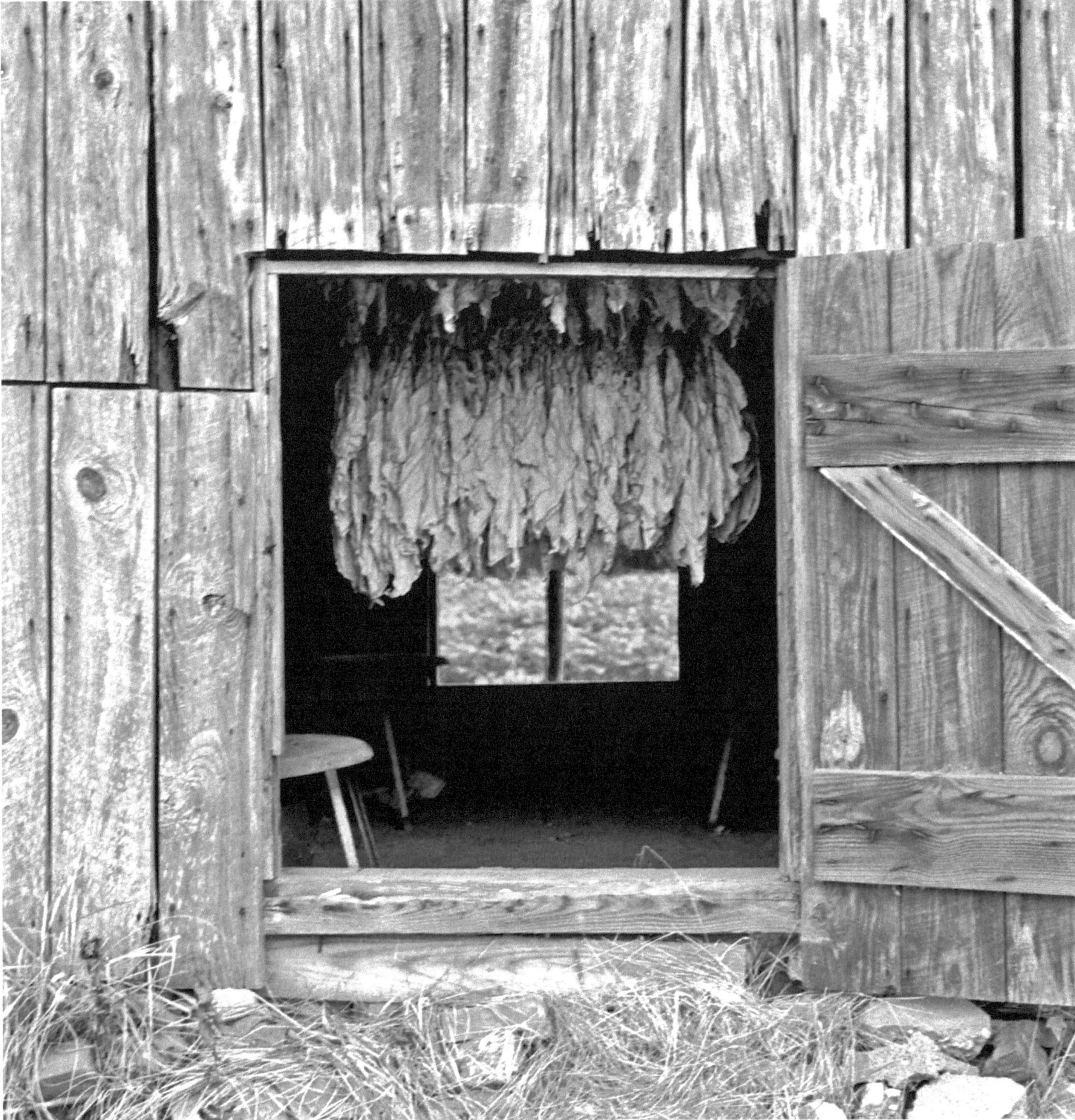

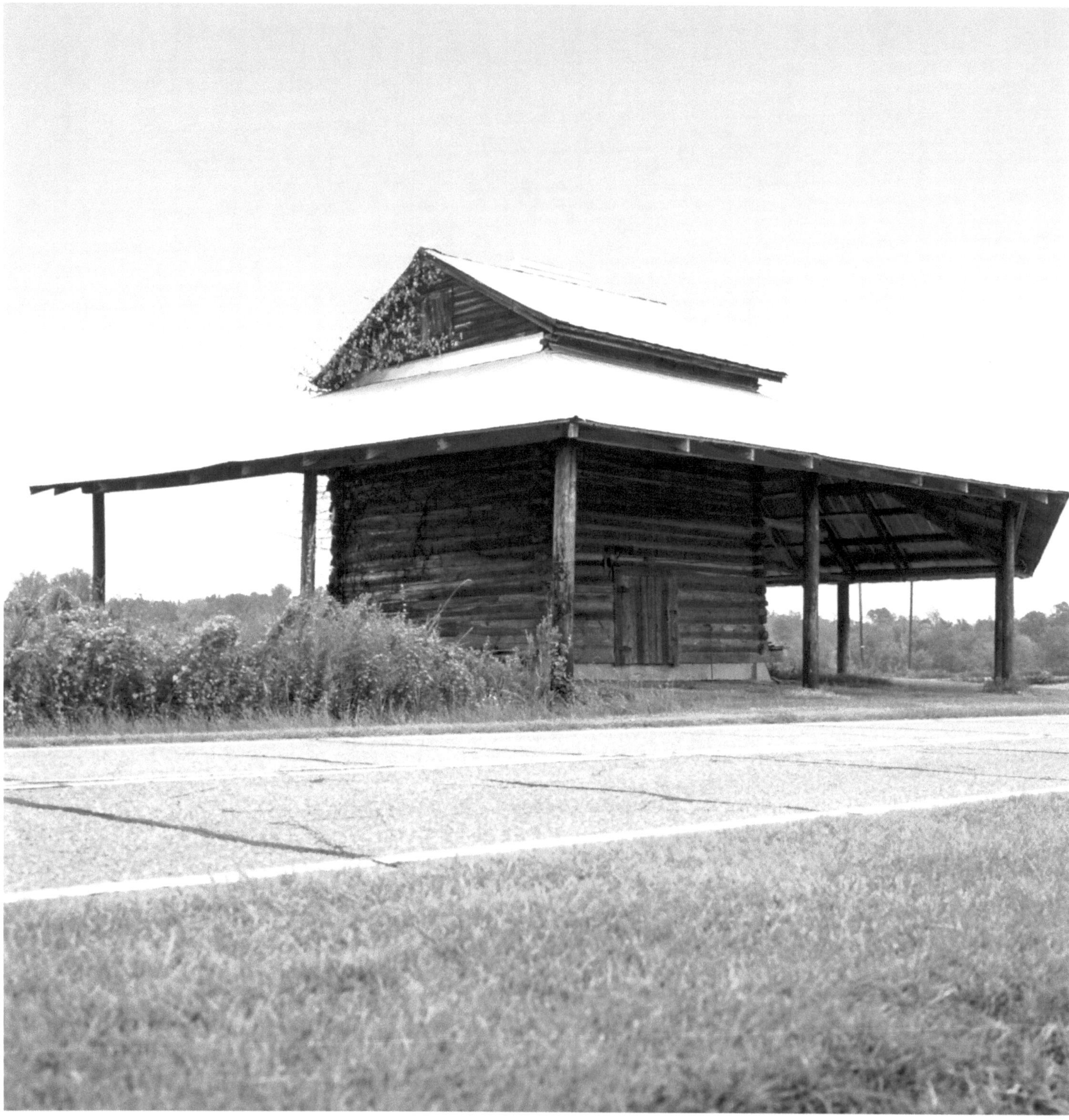

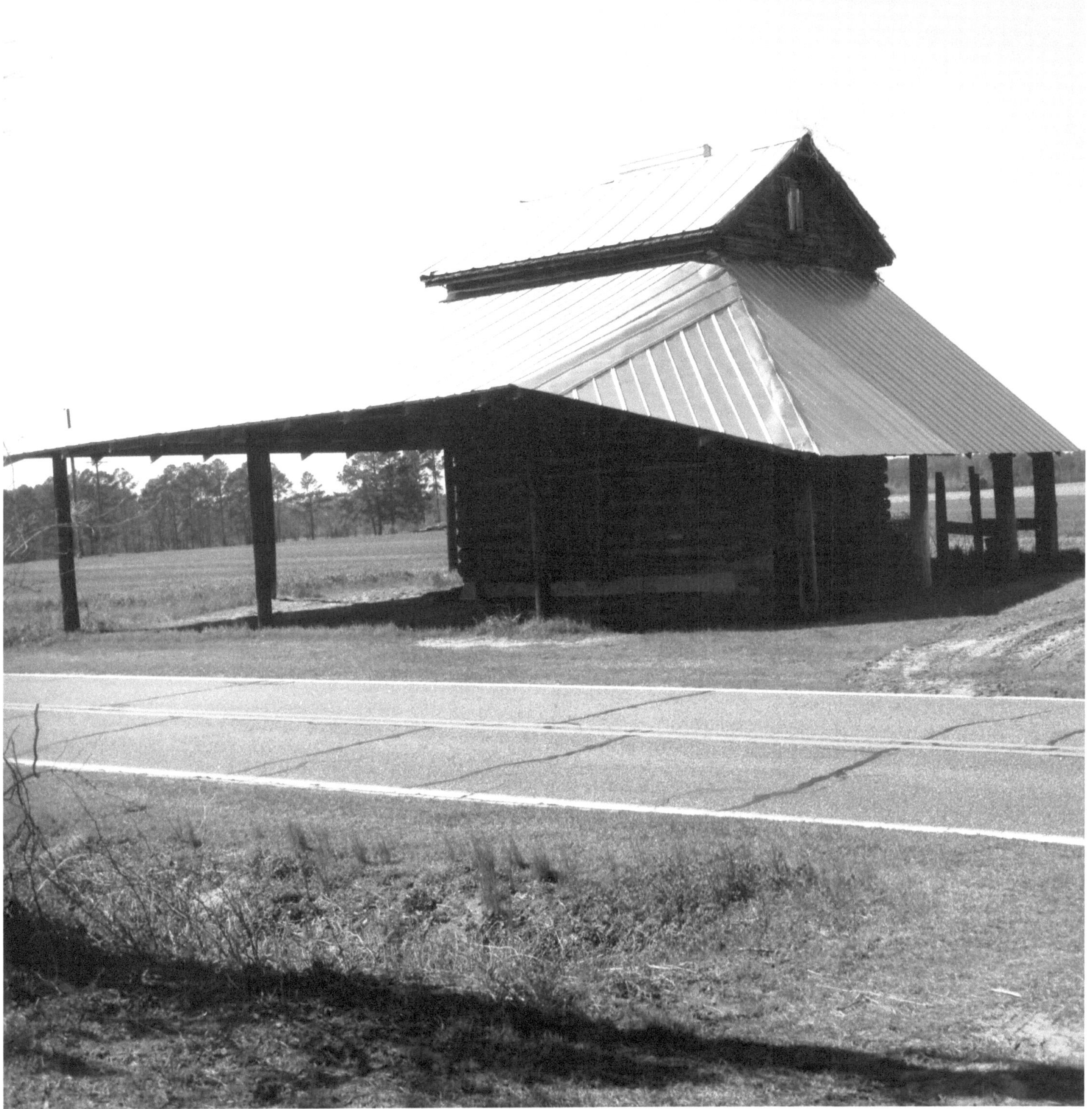

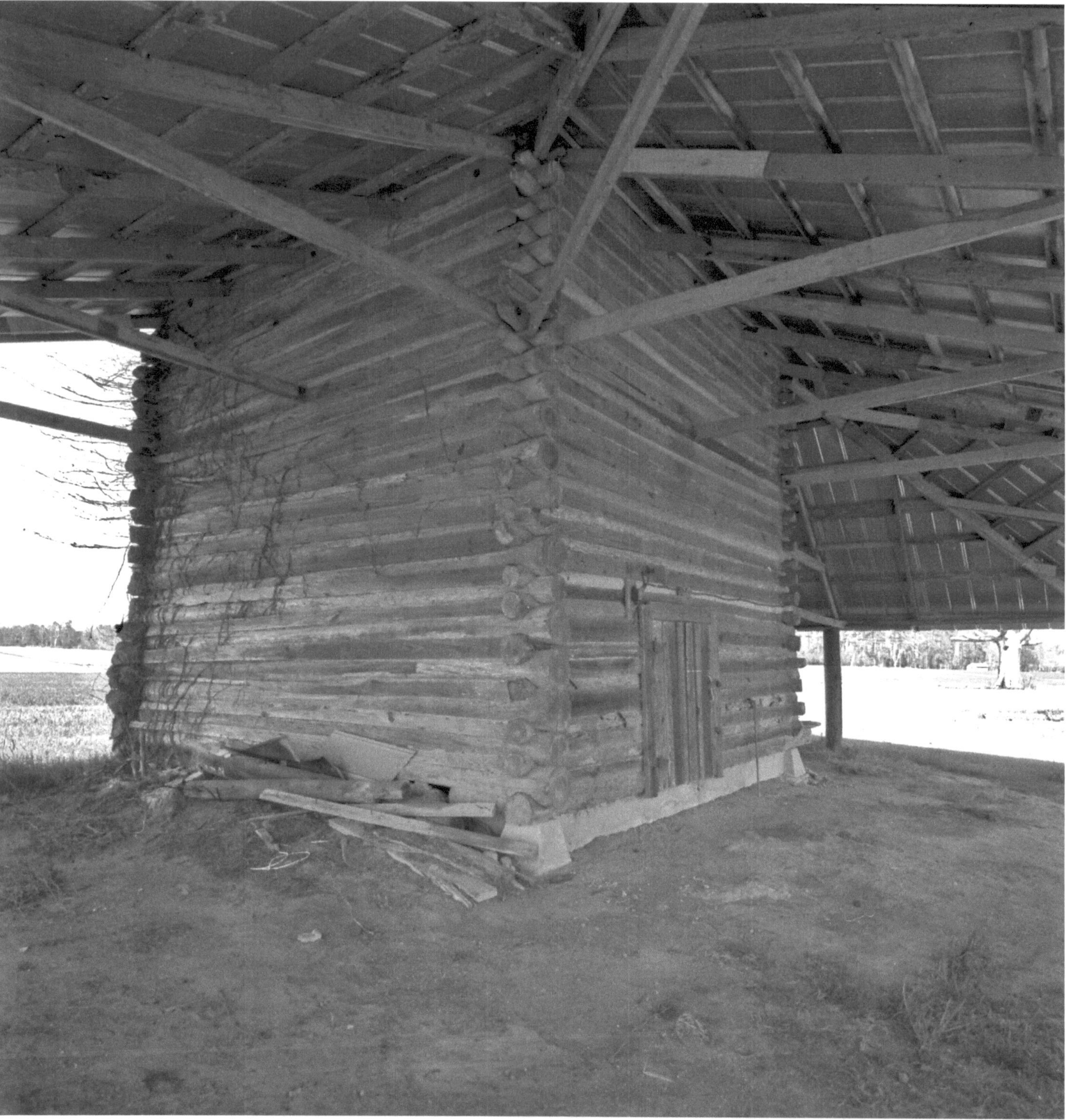

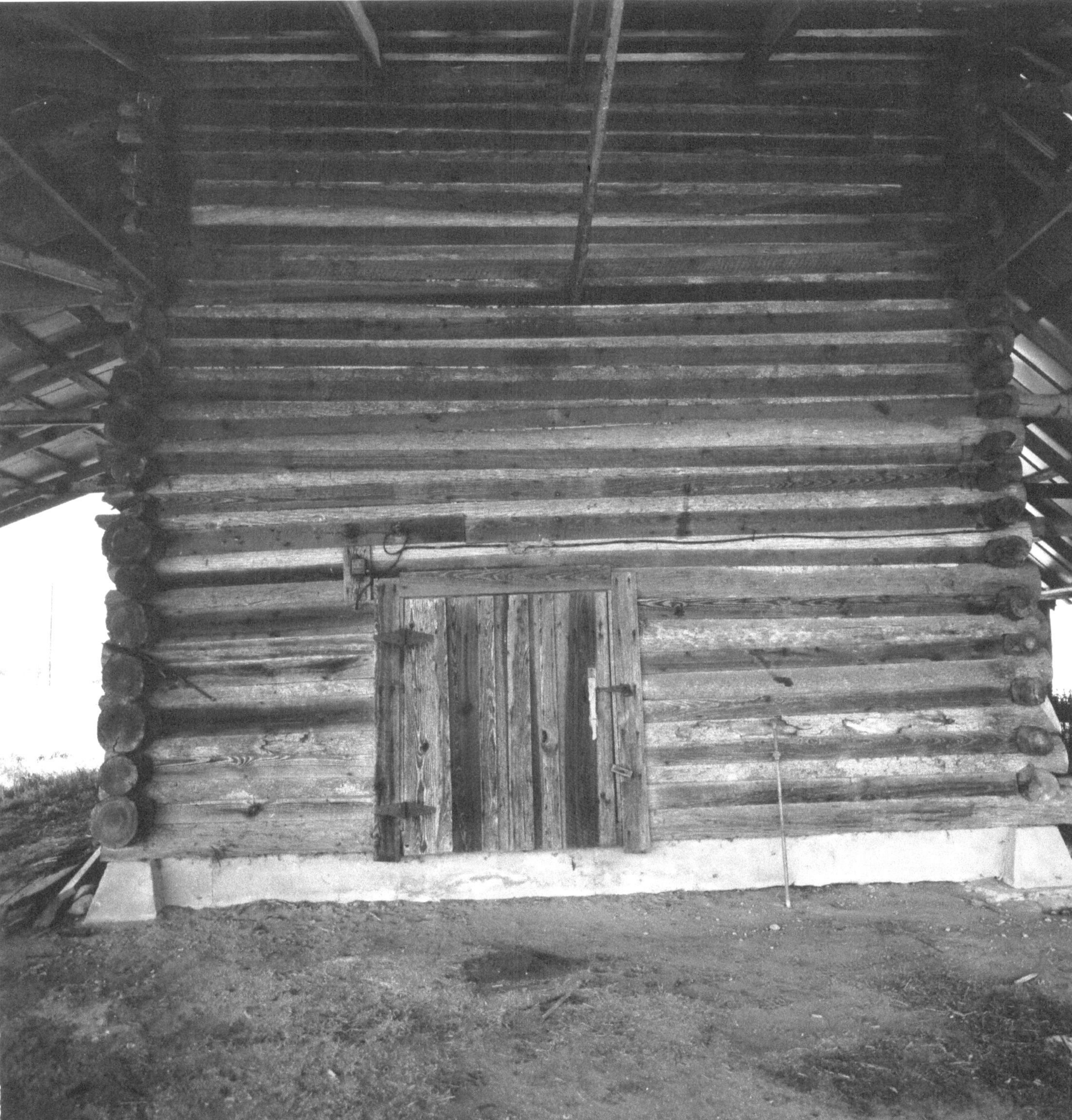

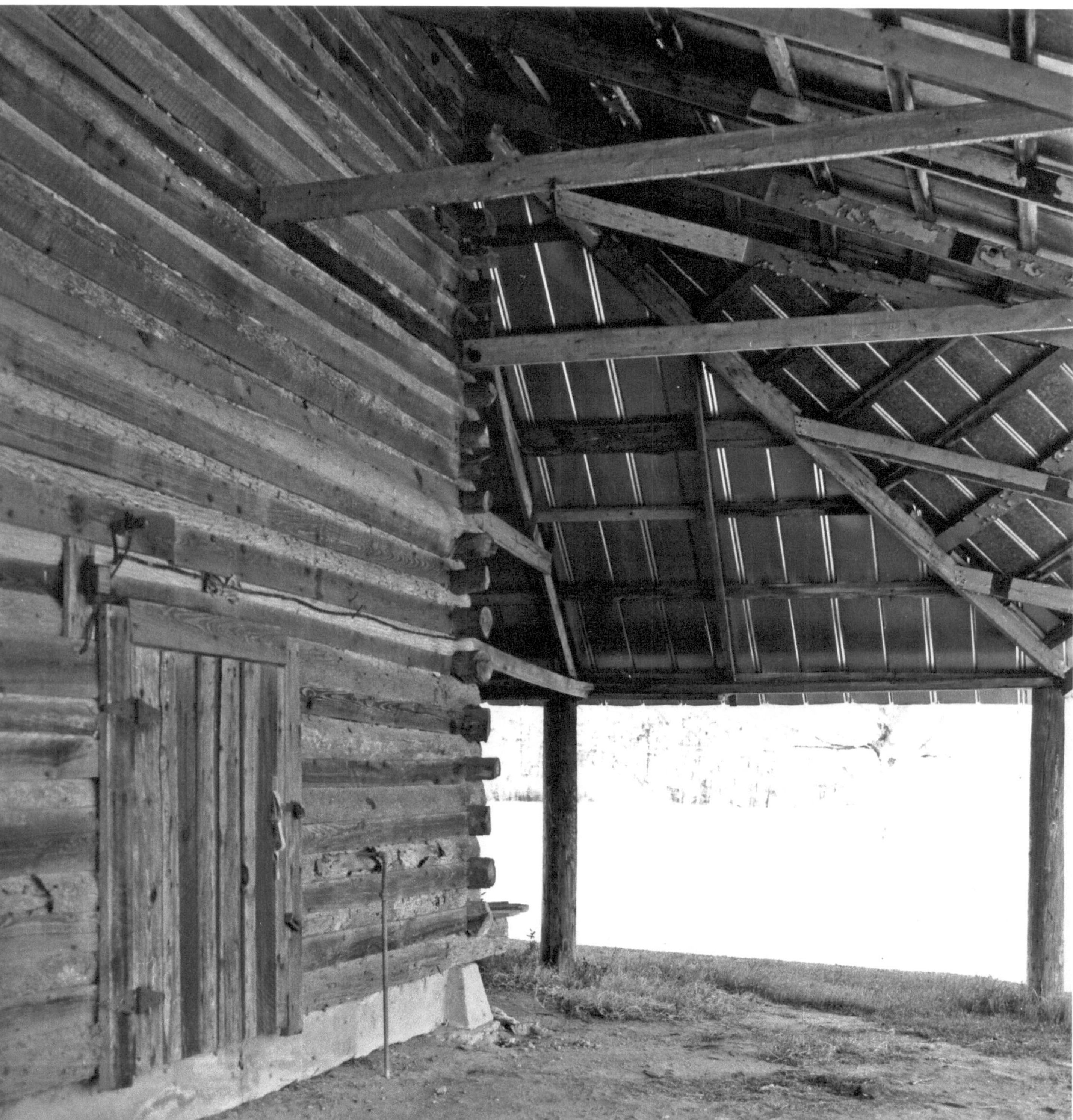

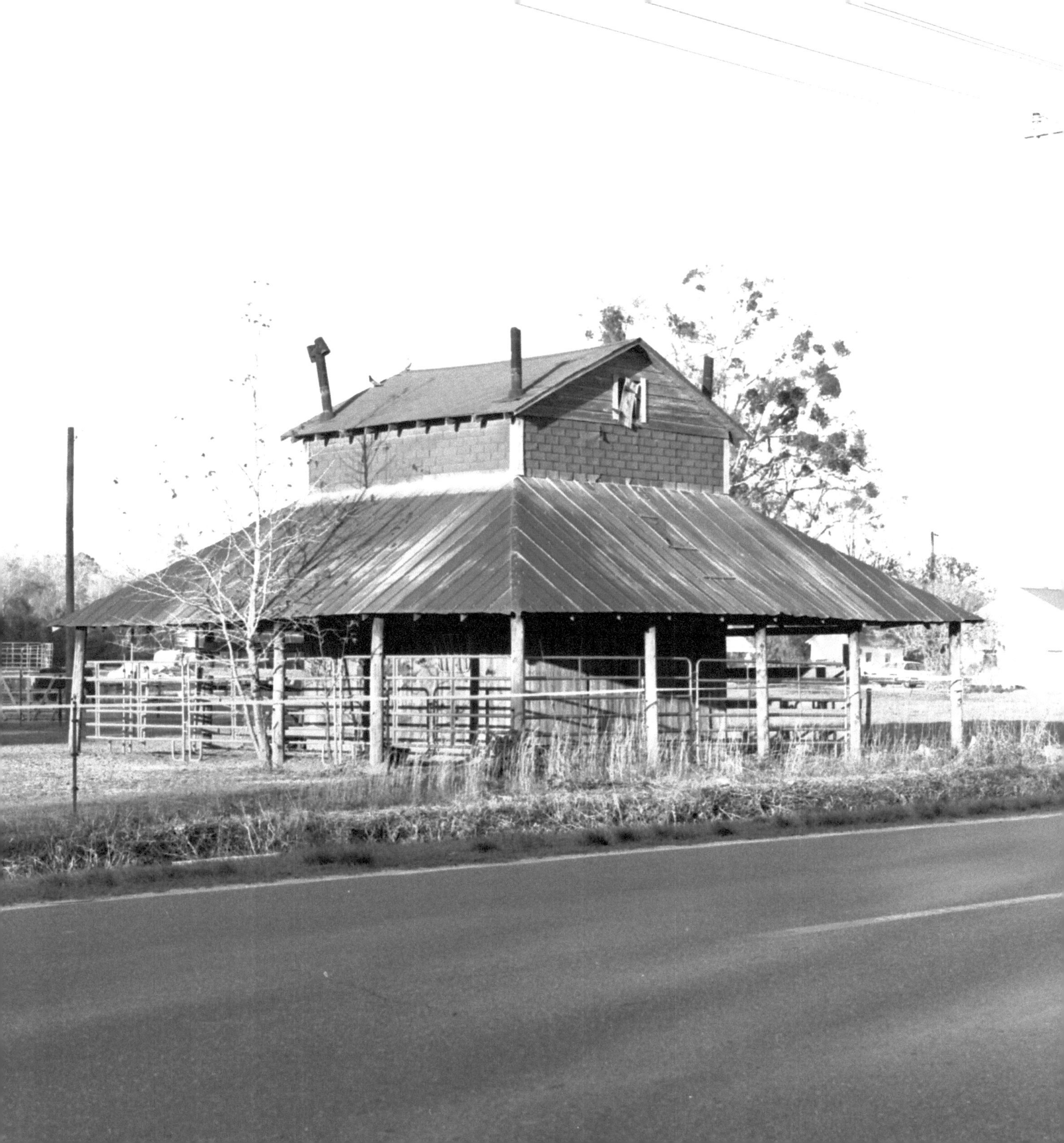

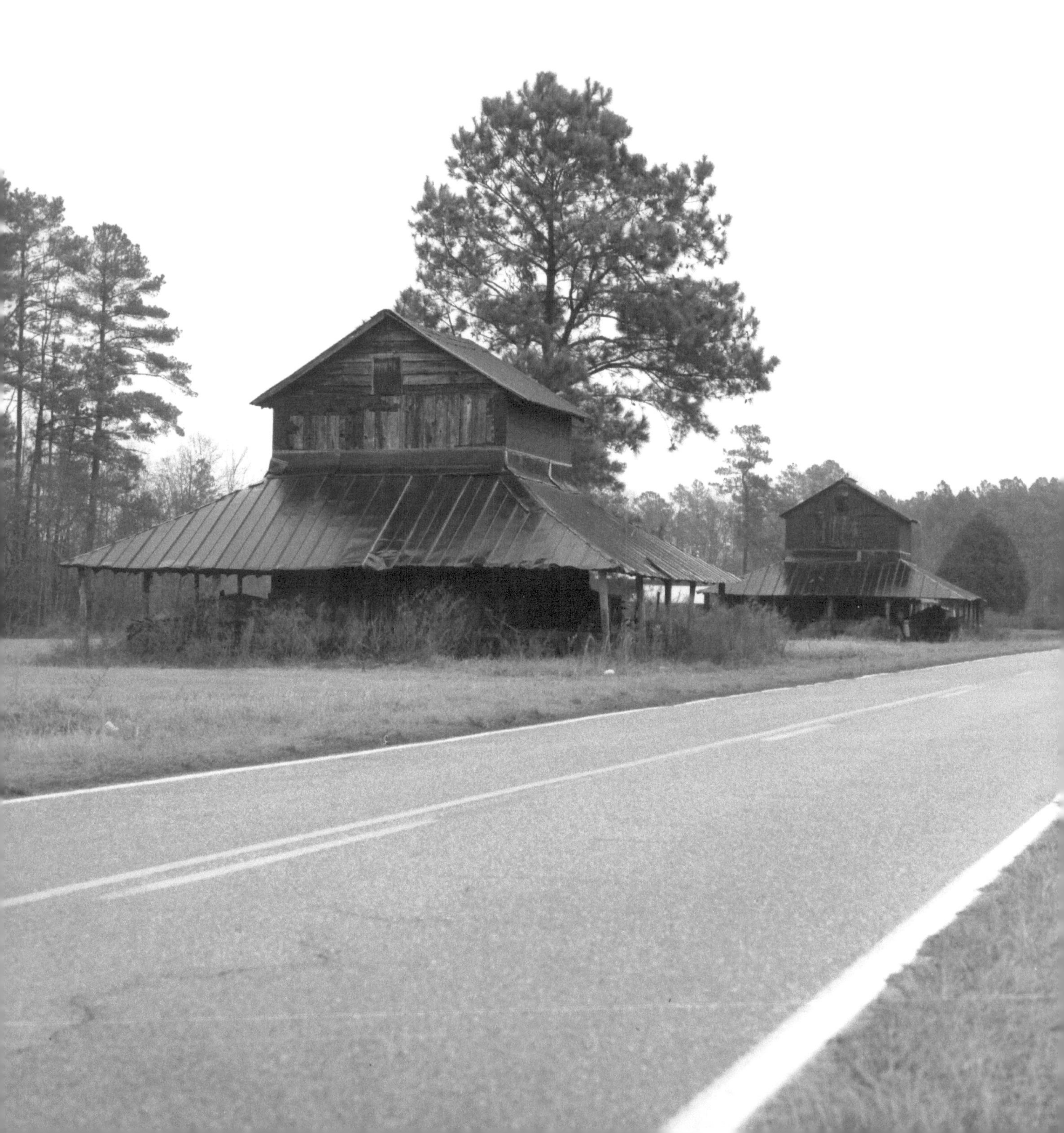

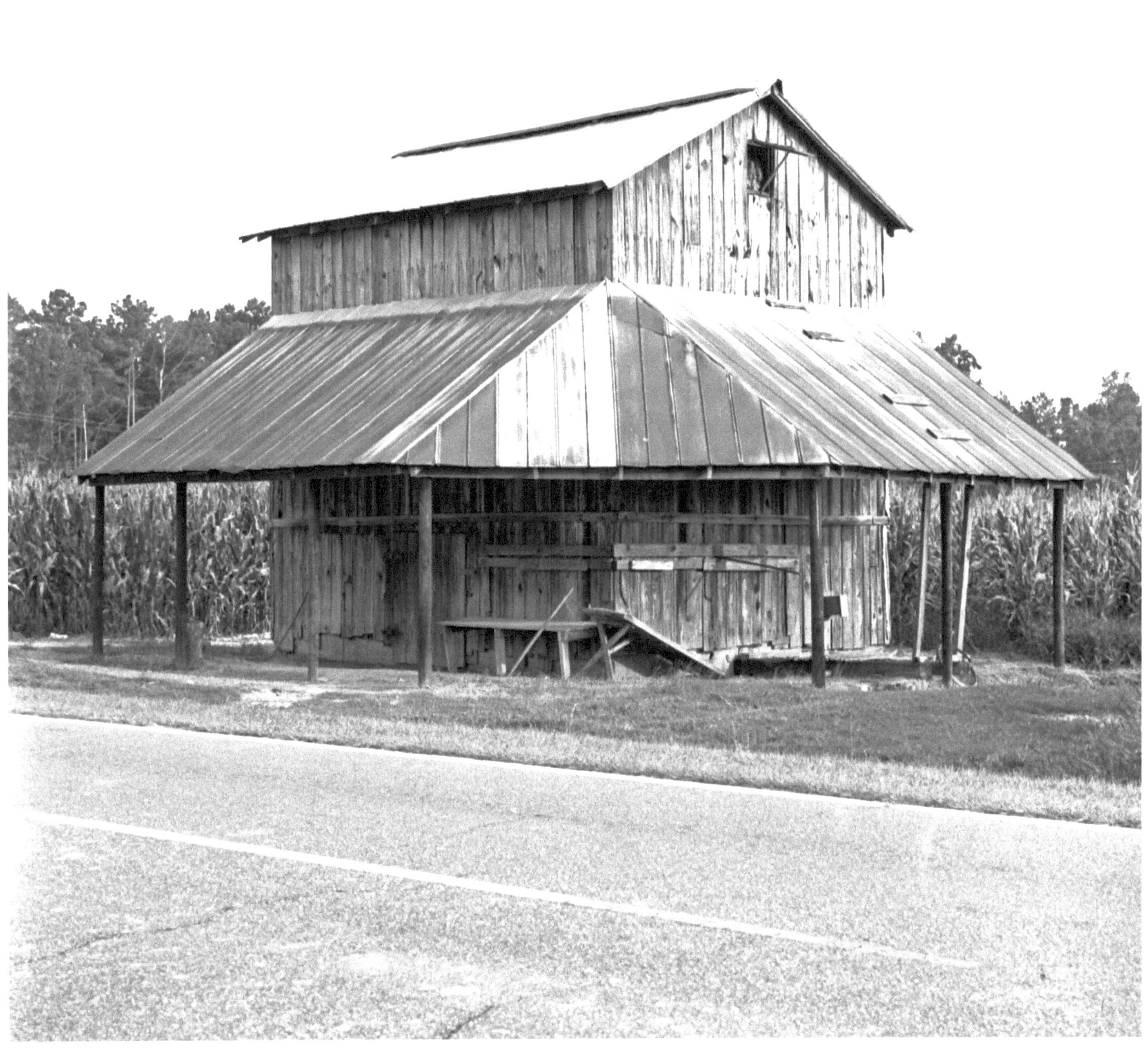

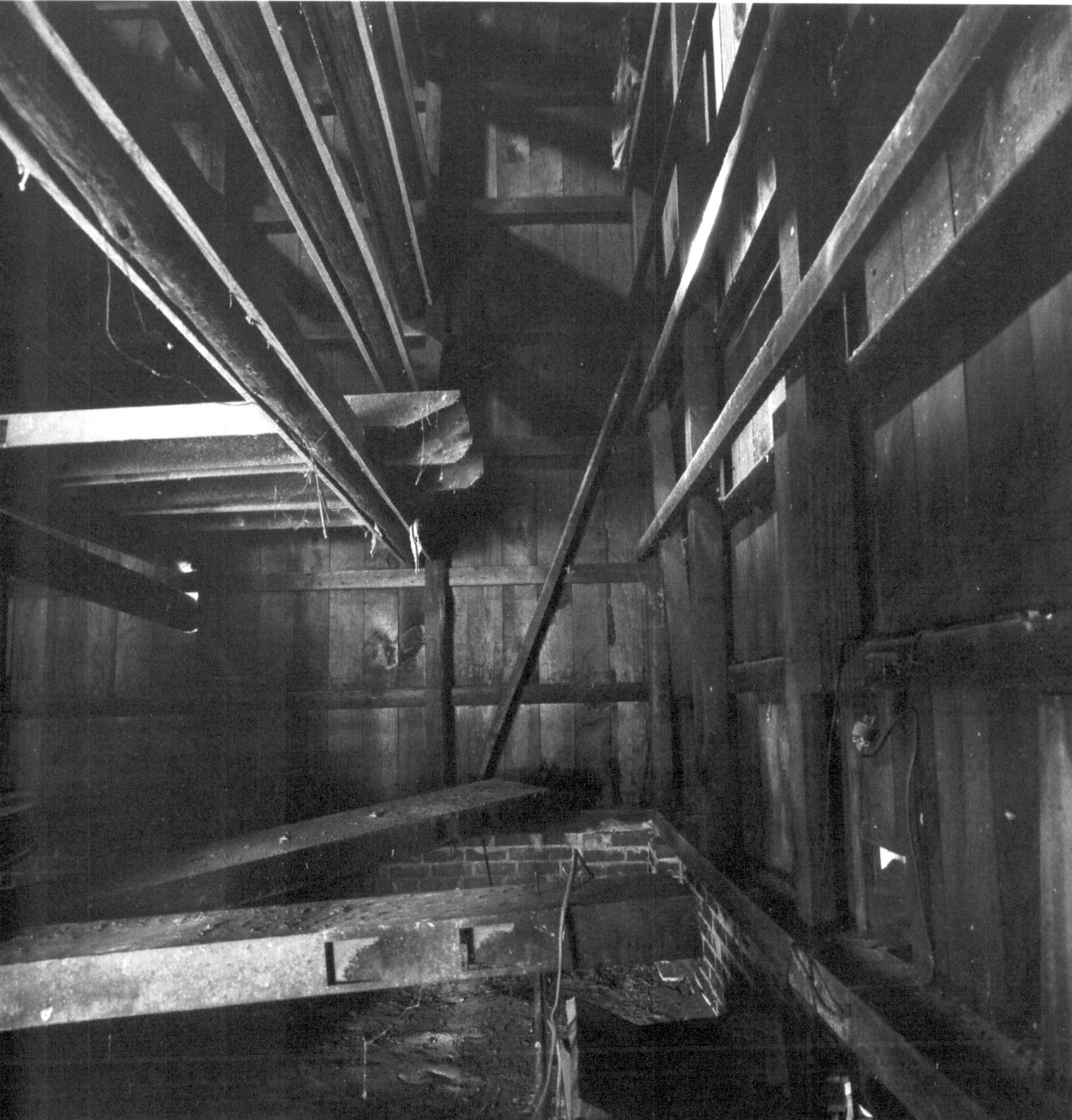

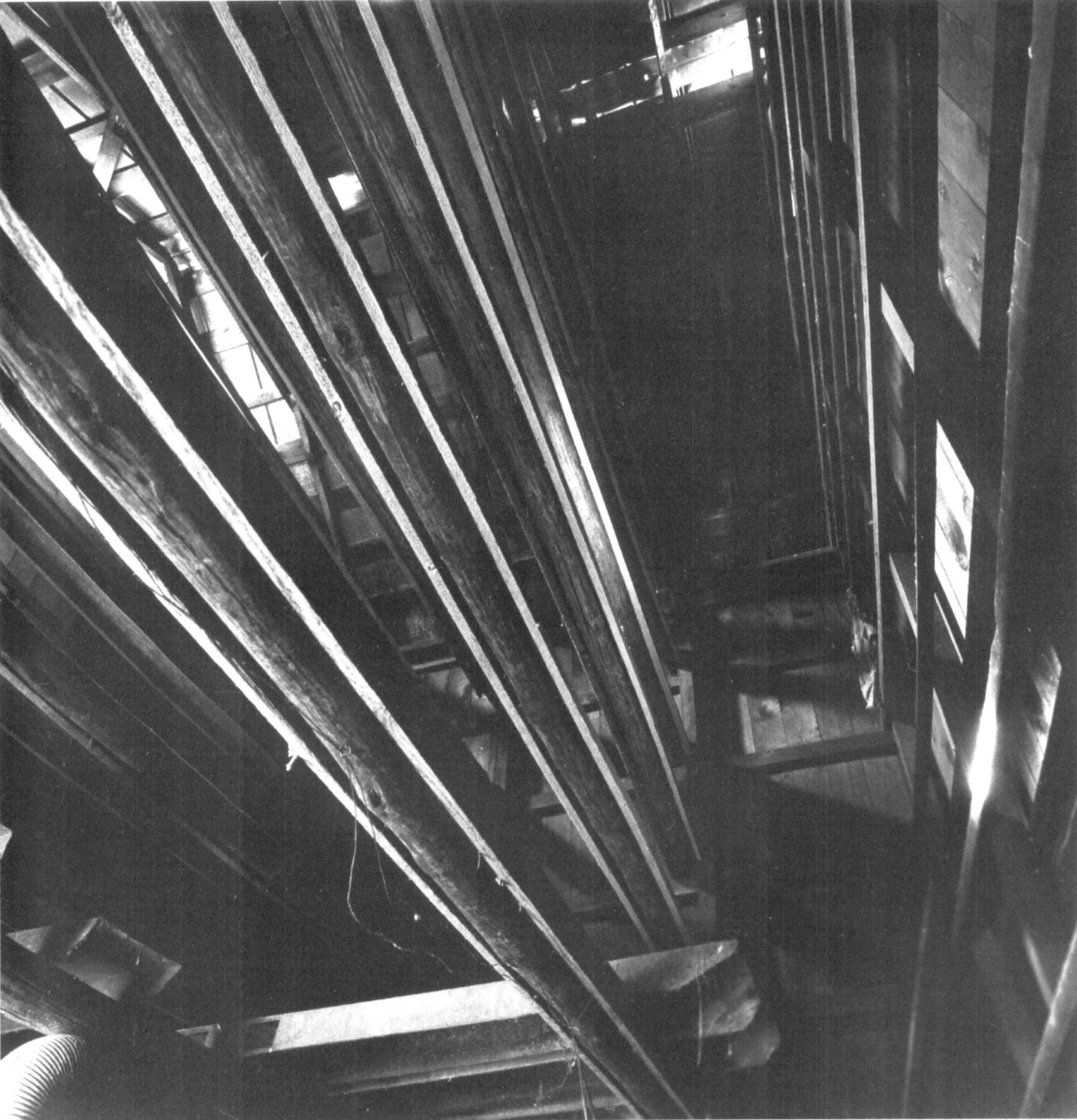

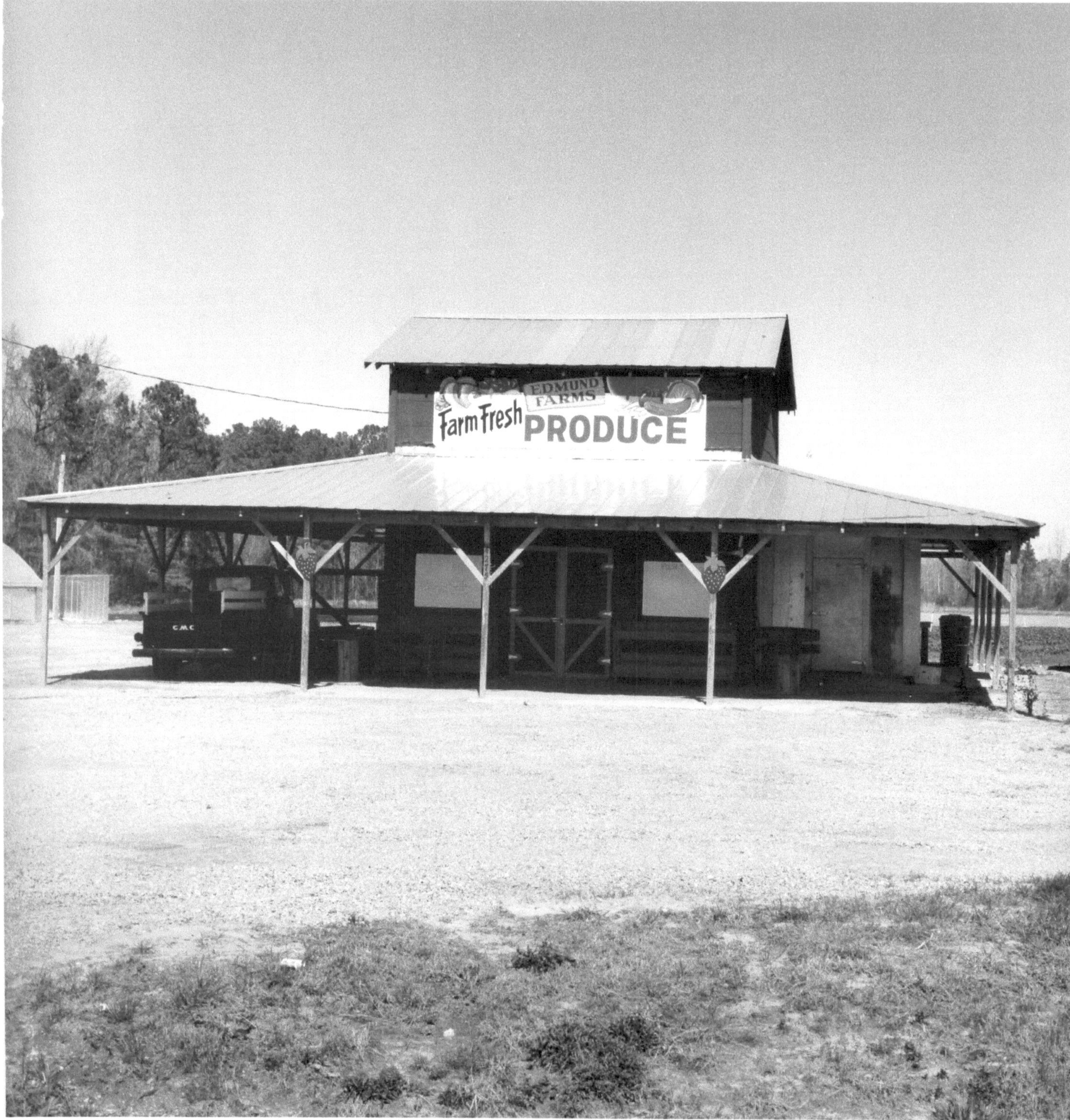
EDMUND FARMS
Farm Fresh PRODUCE
GMC

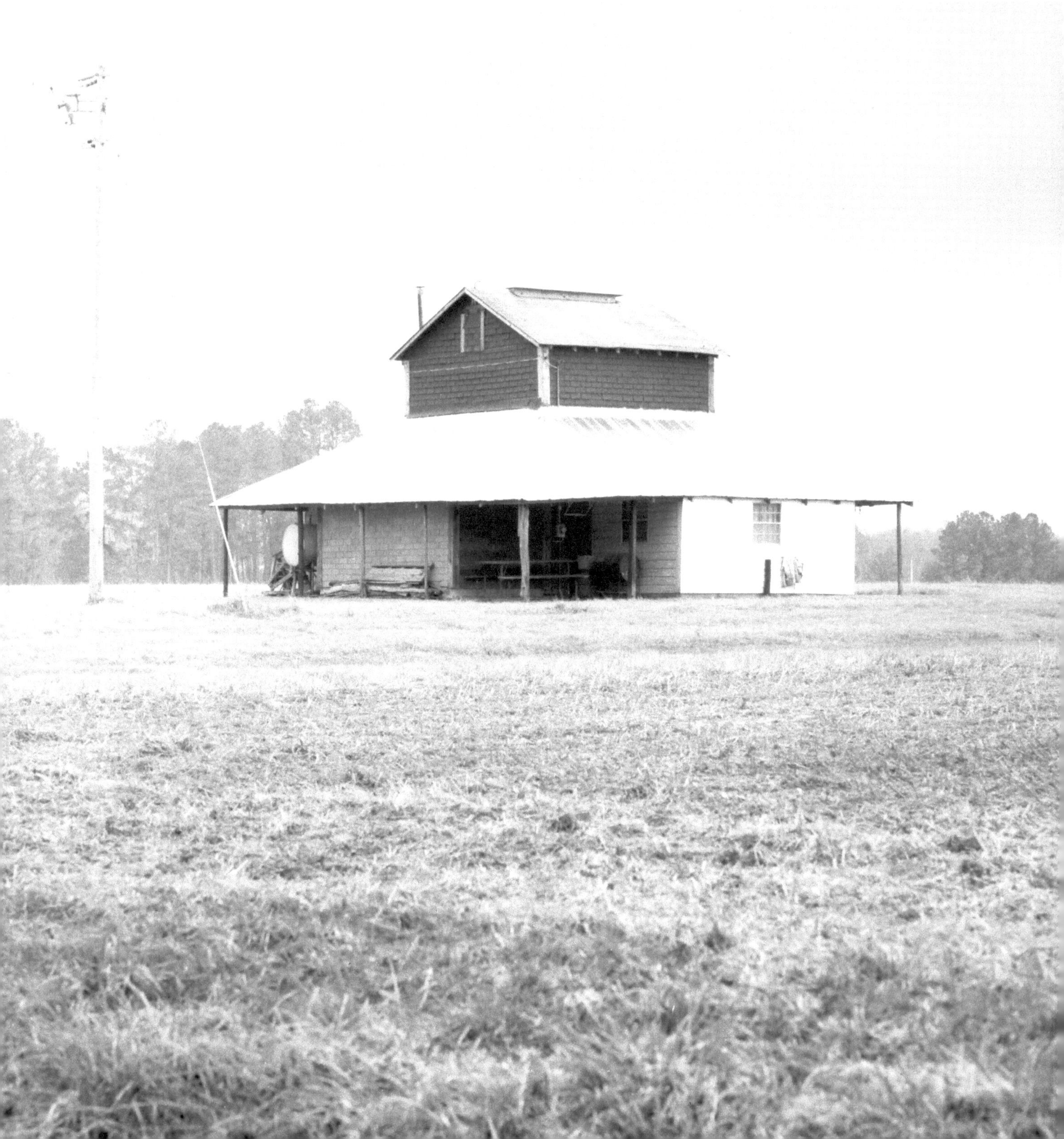

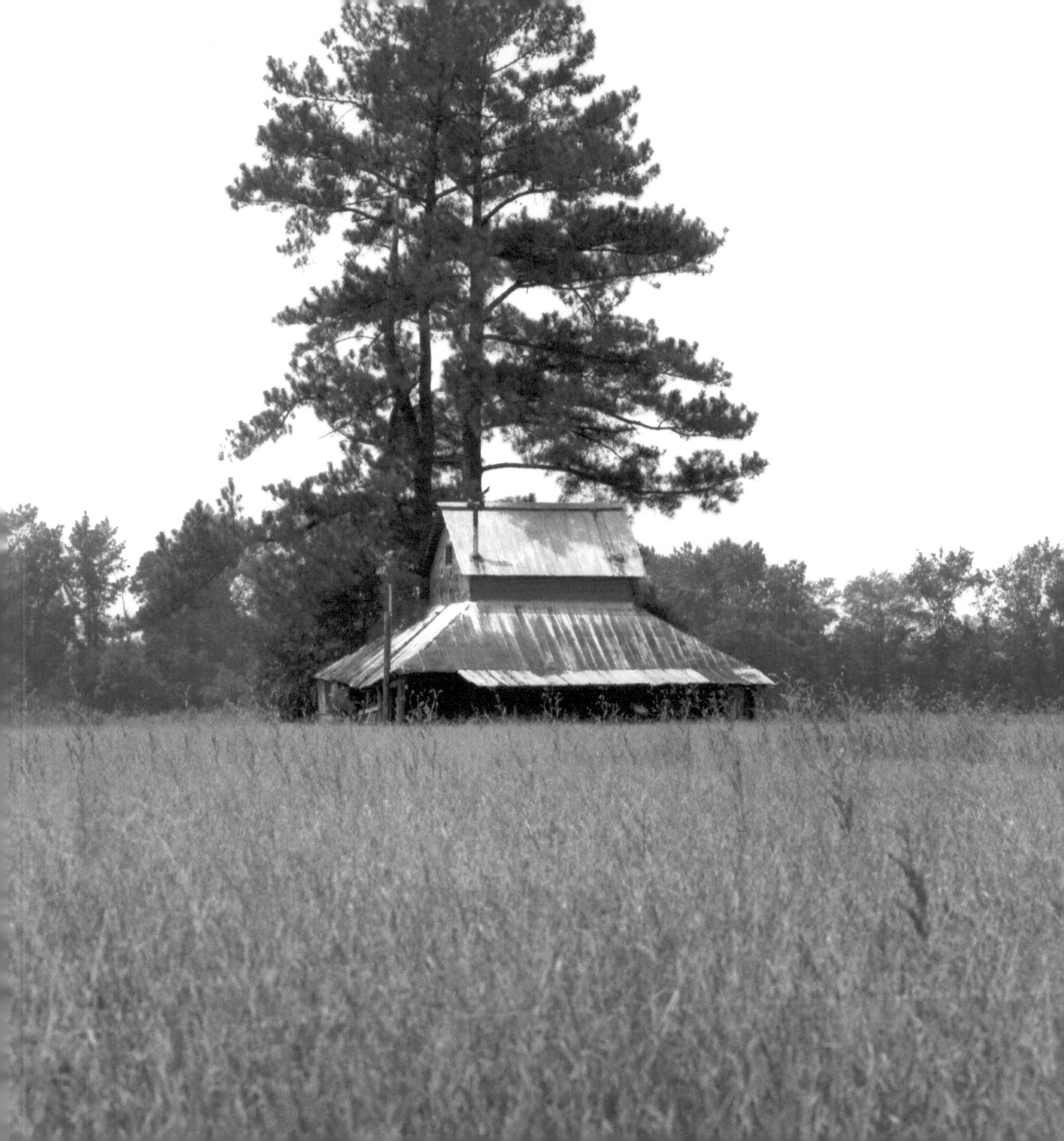

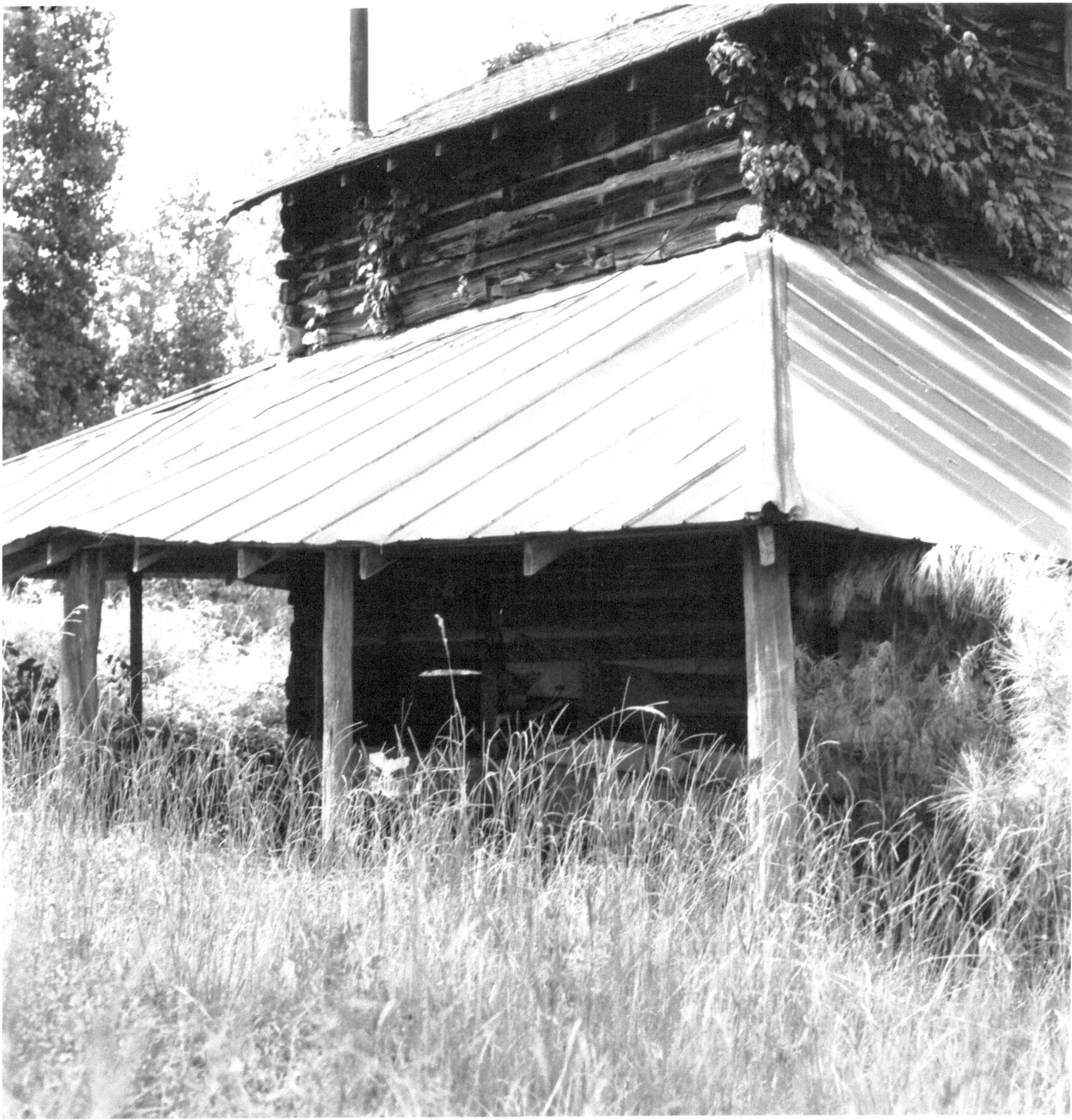

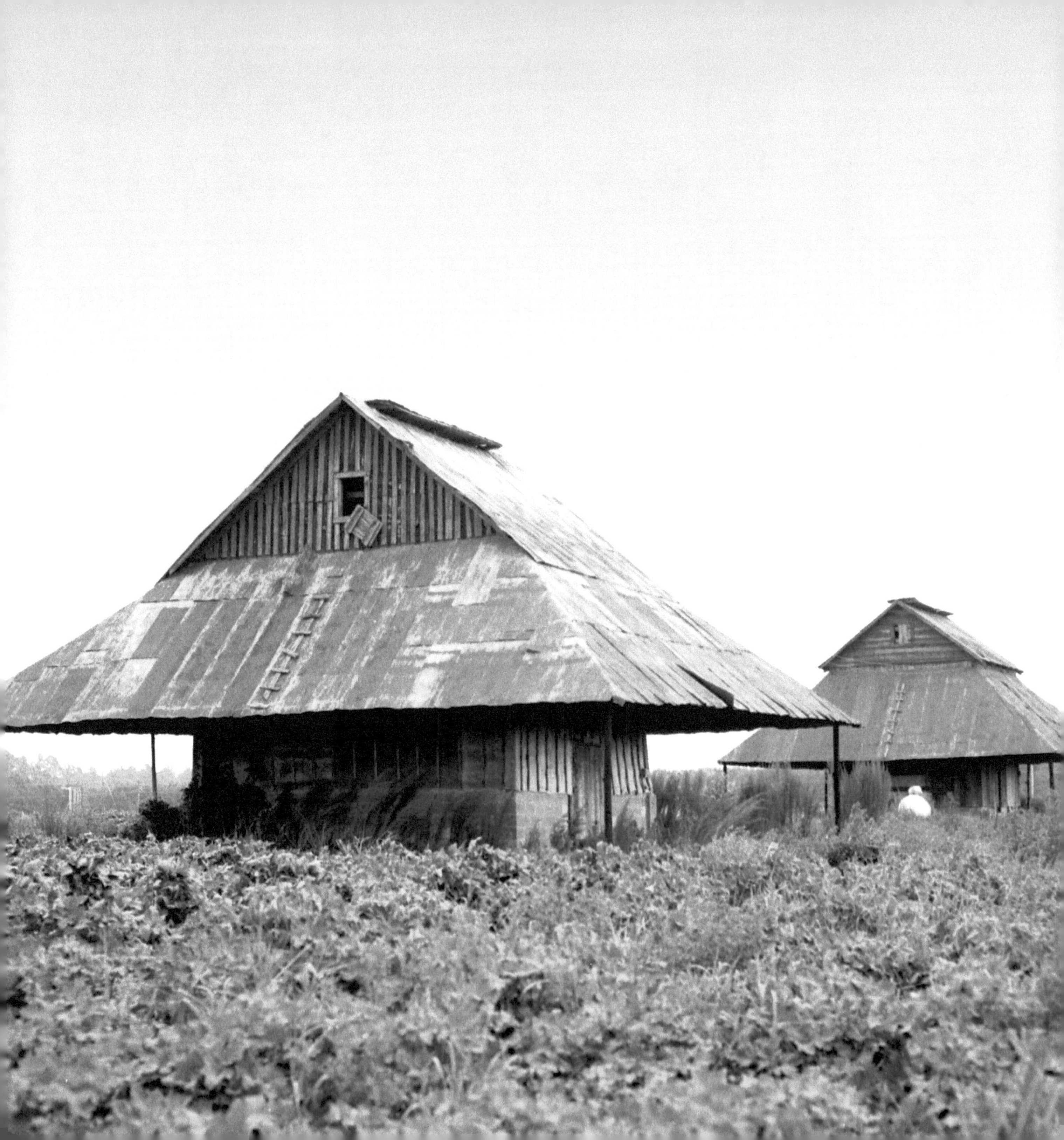

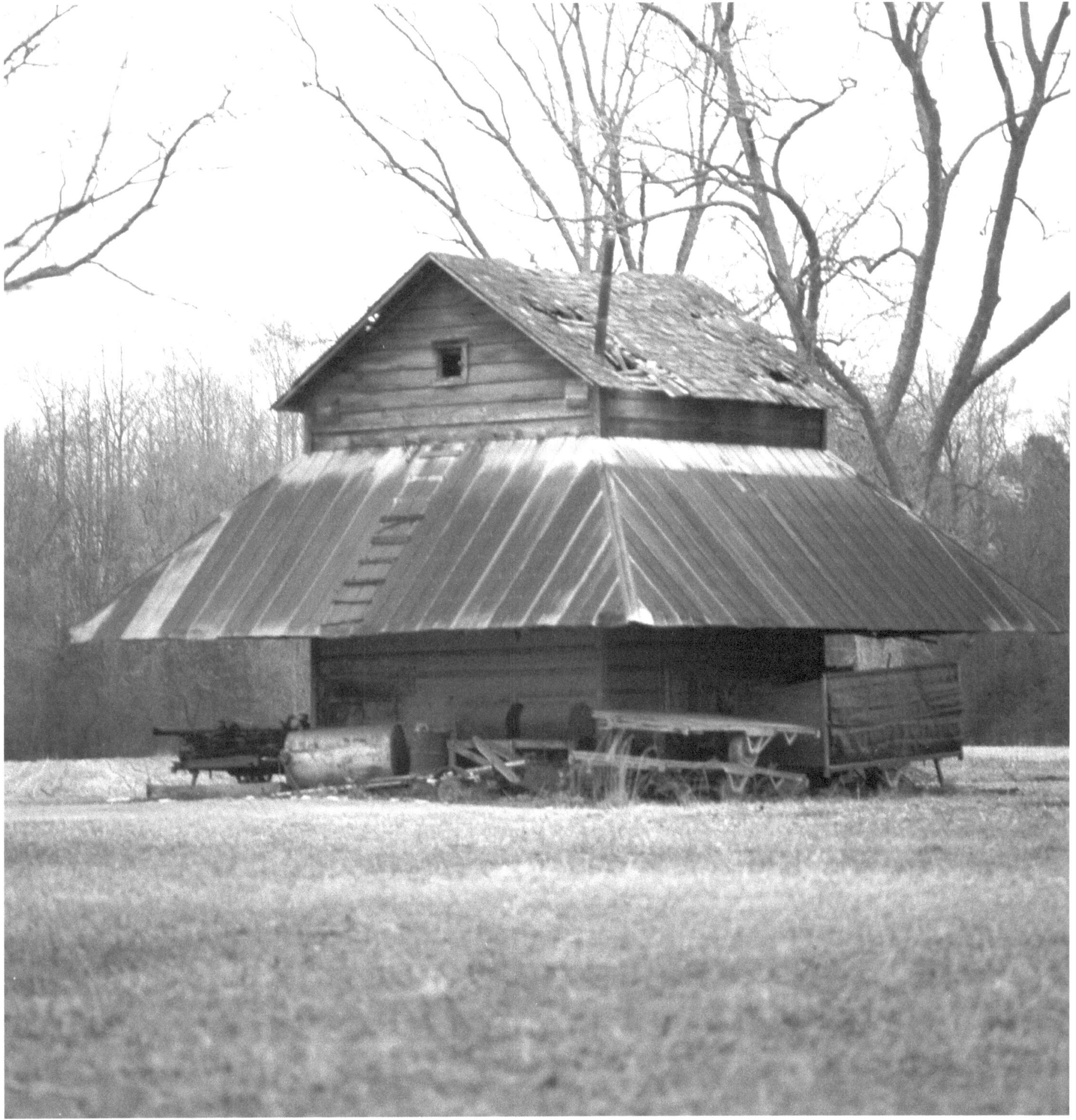

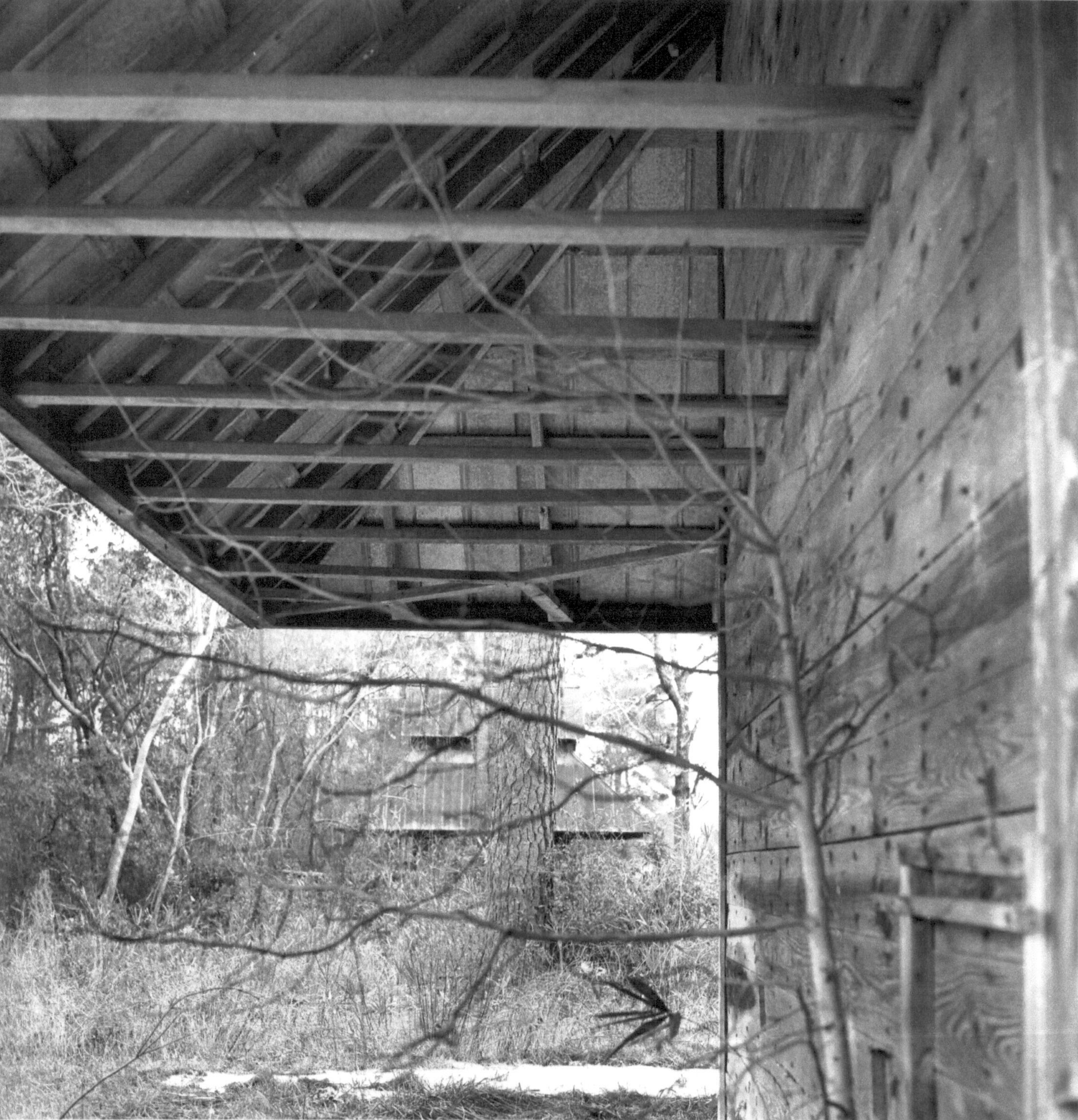

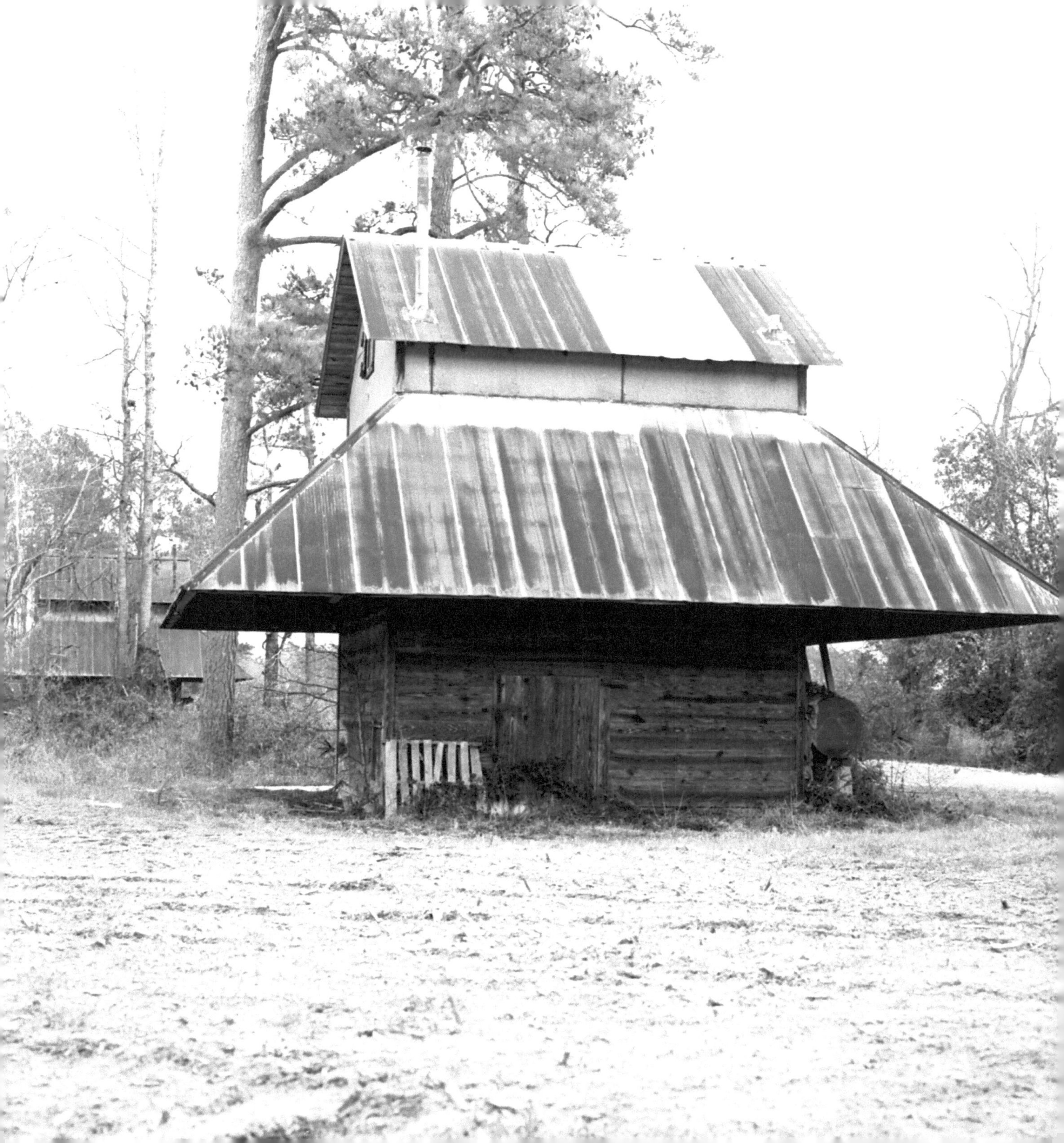

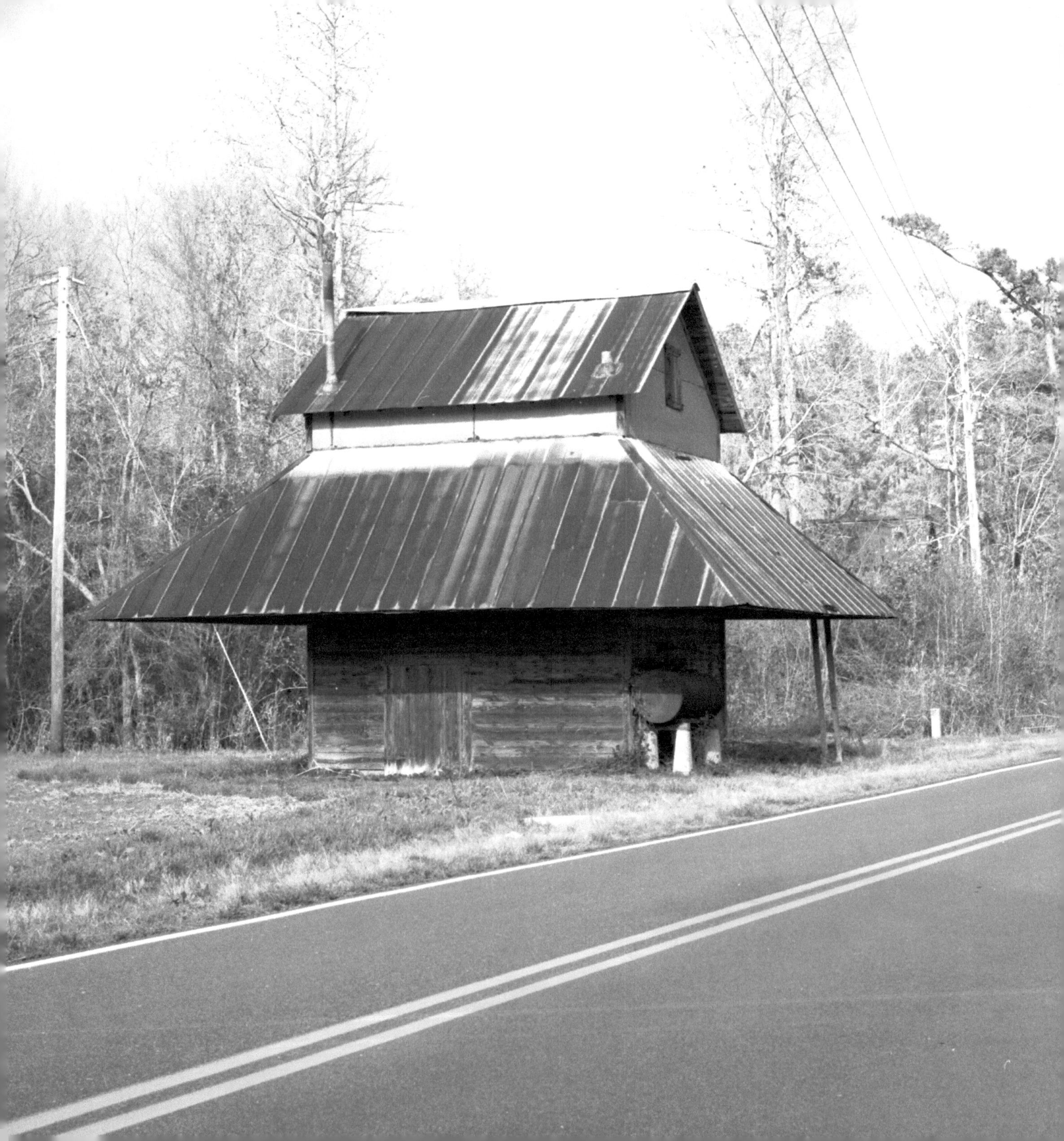

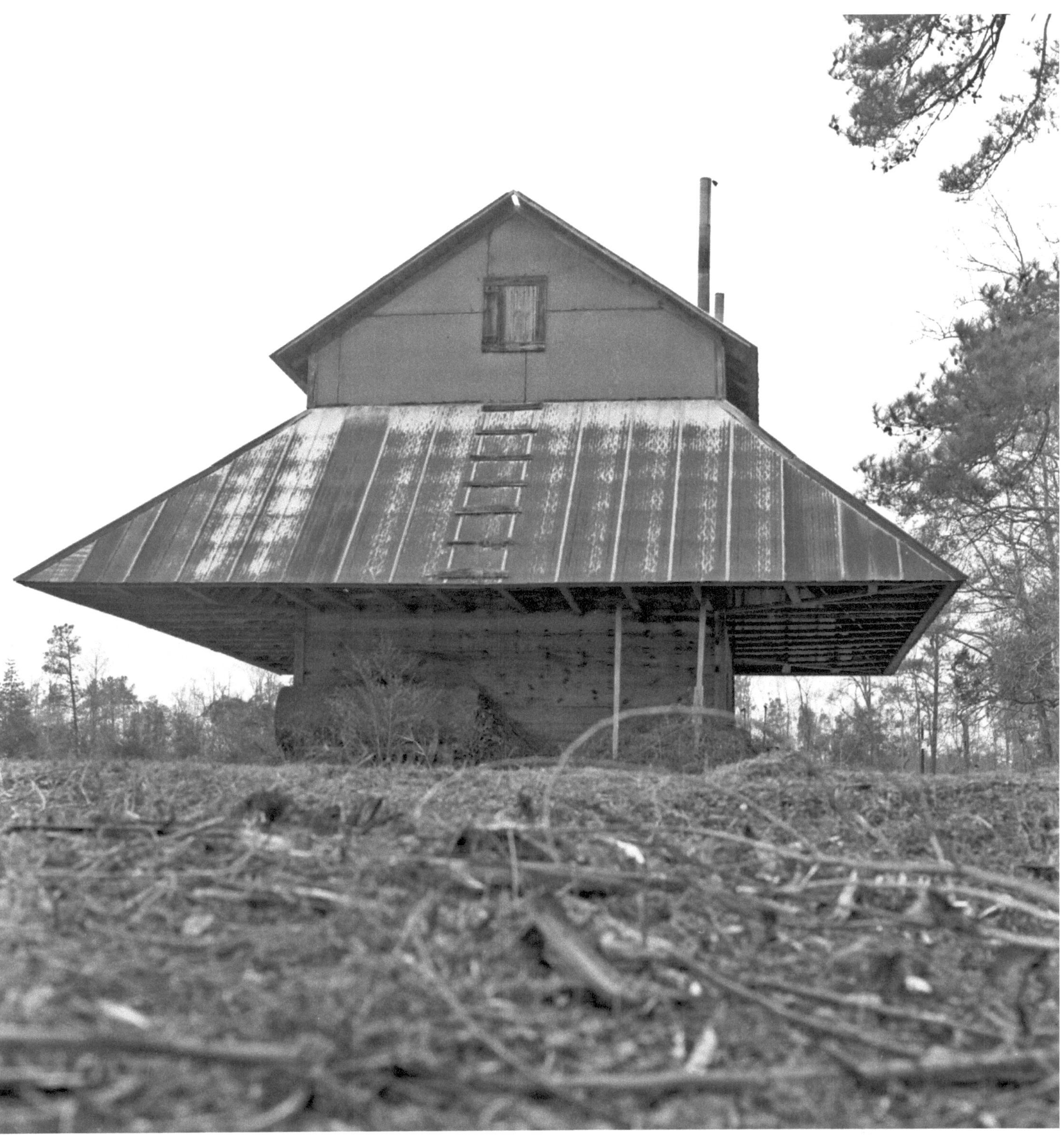

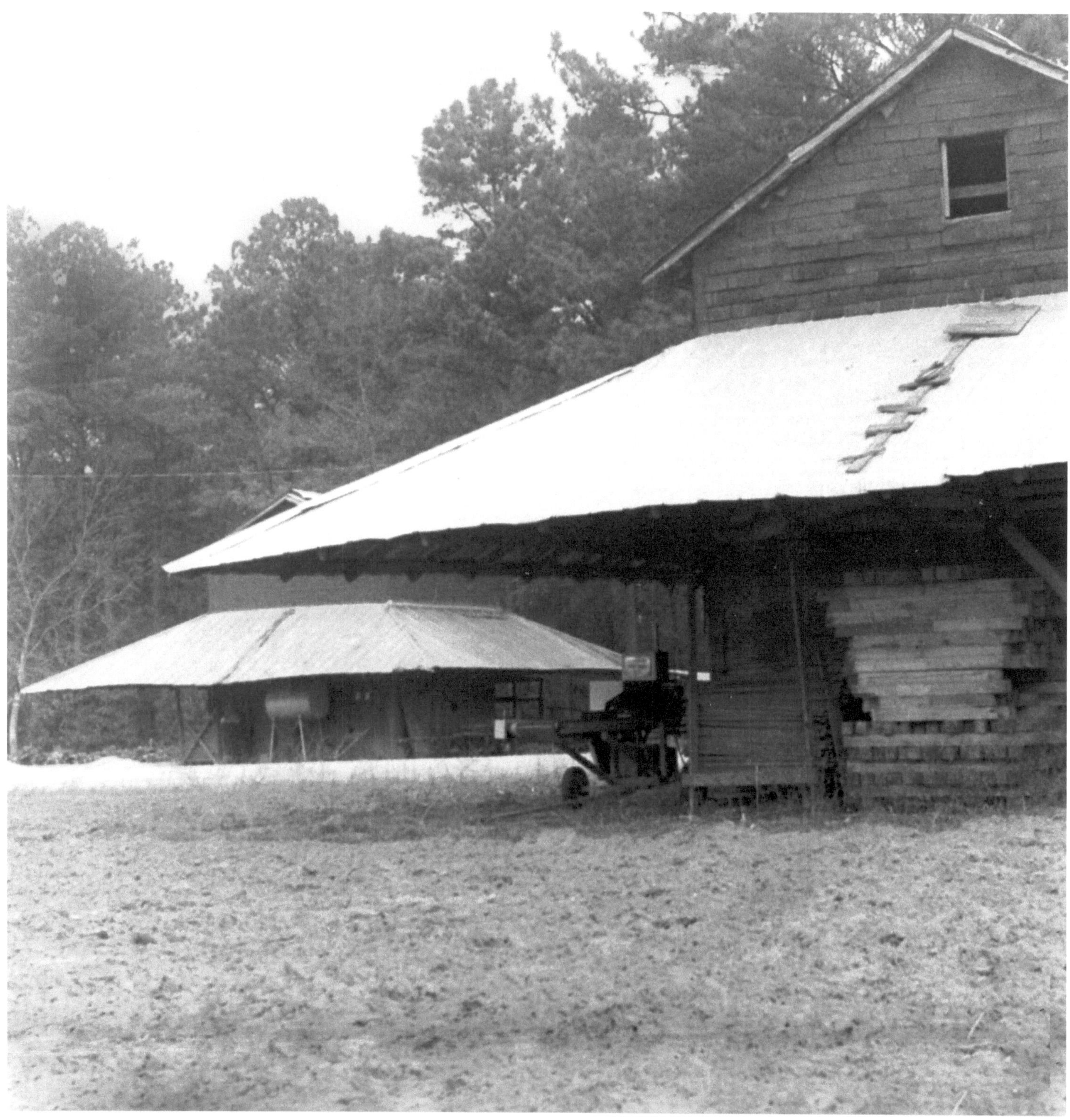

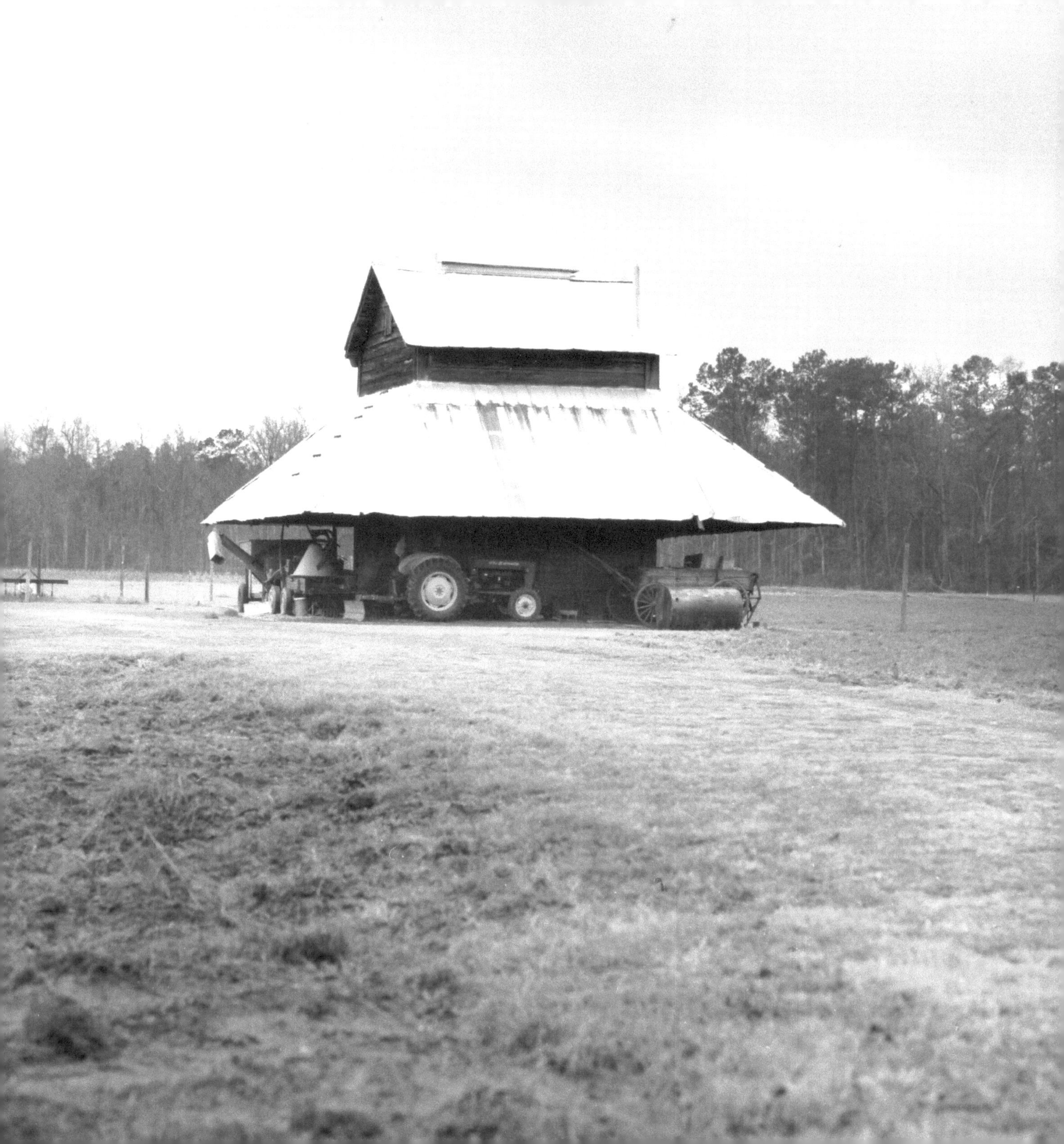

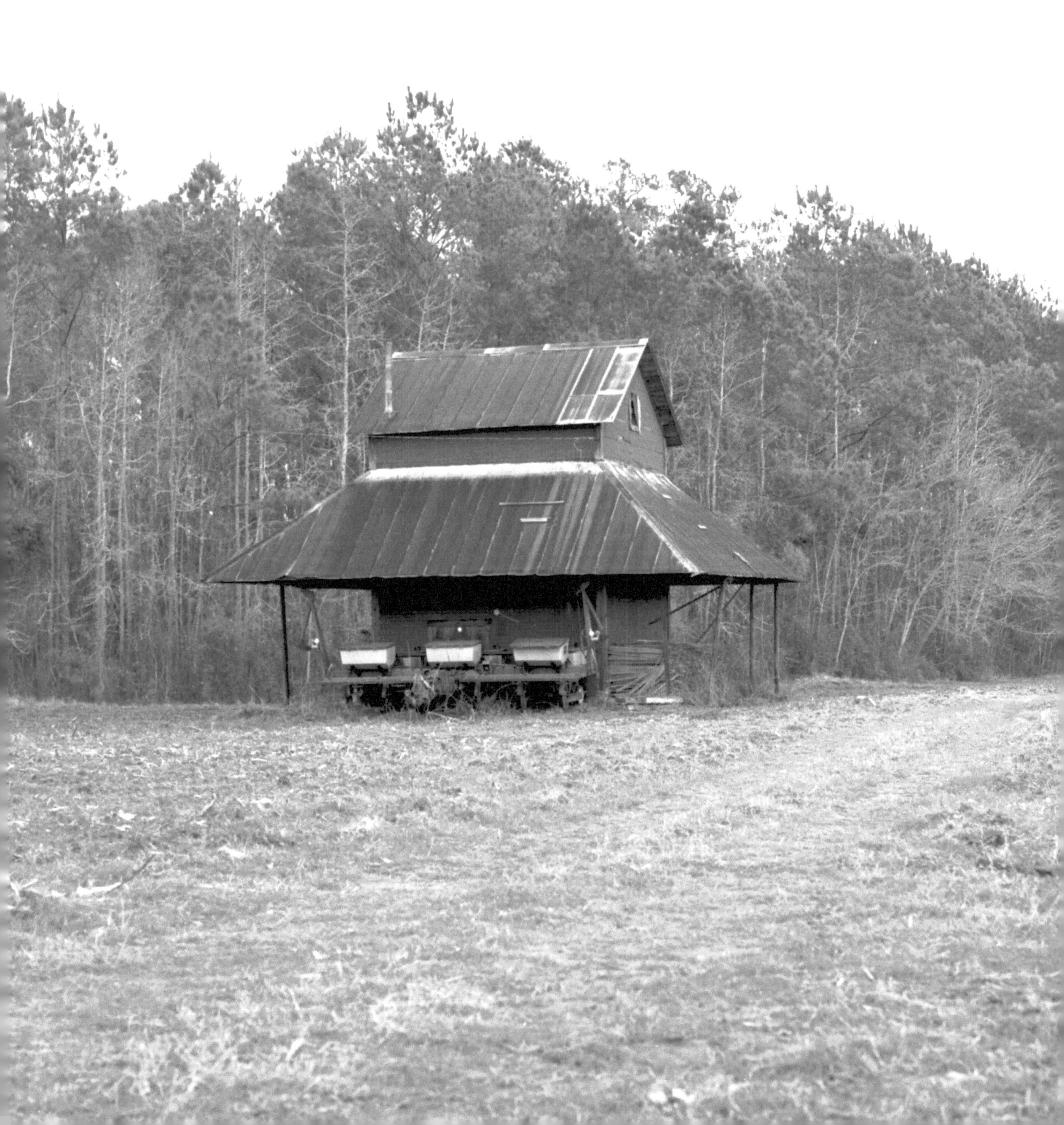

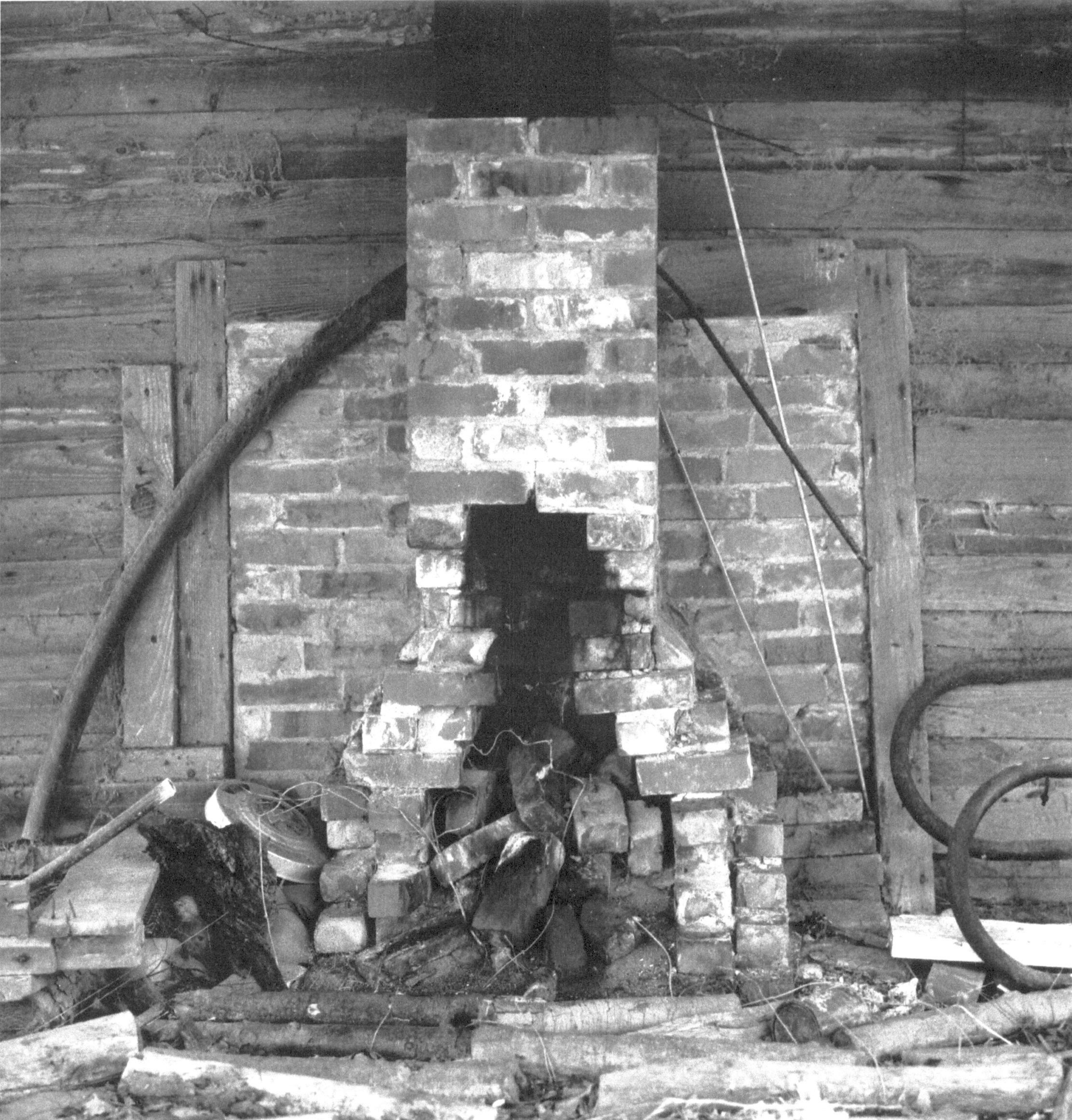

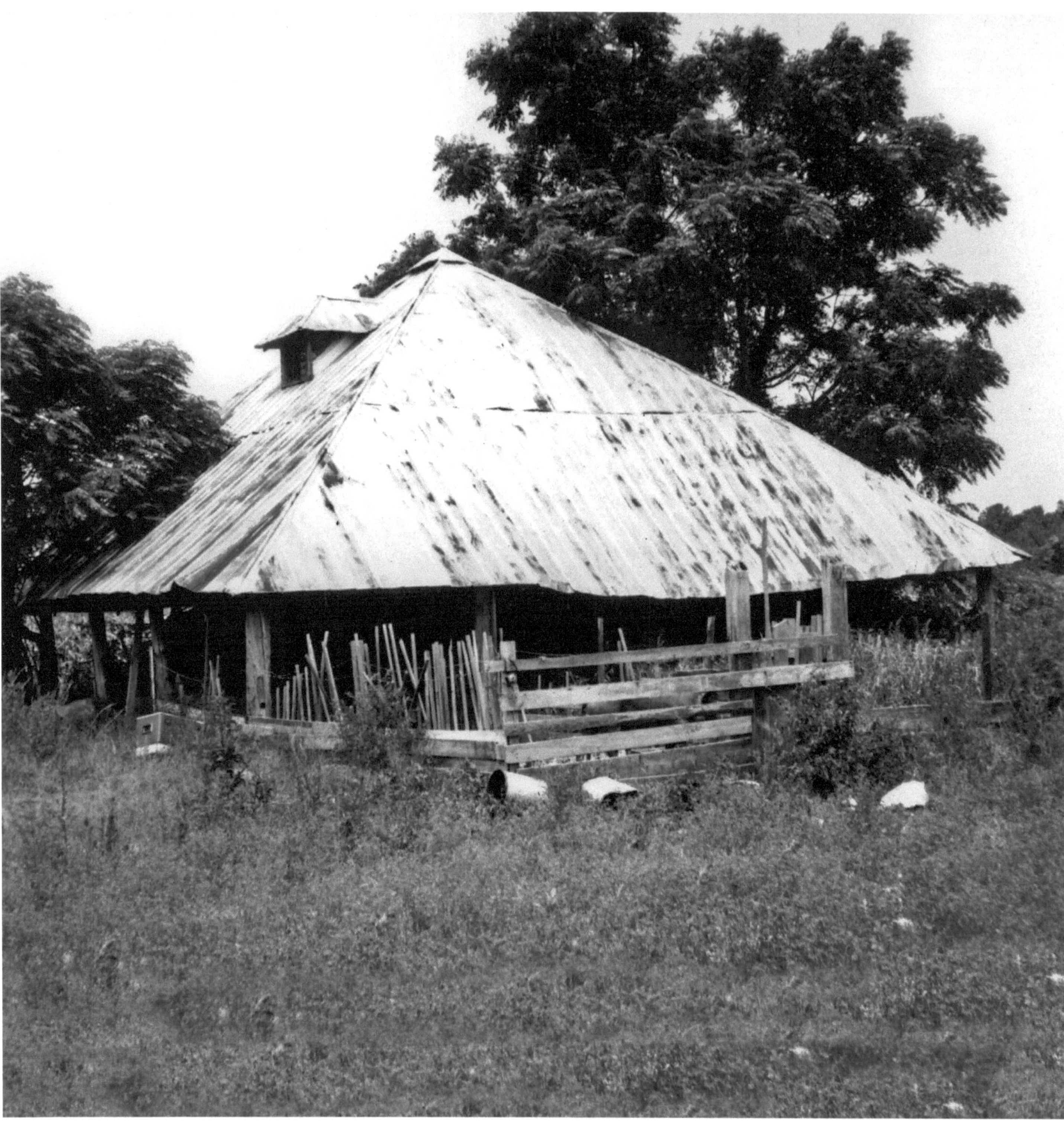

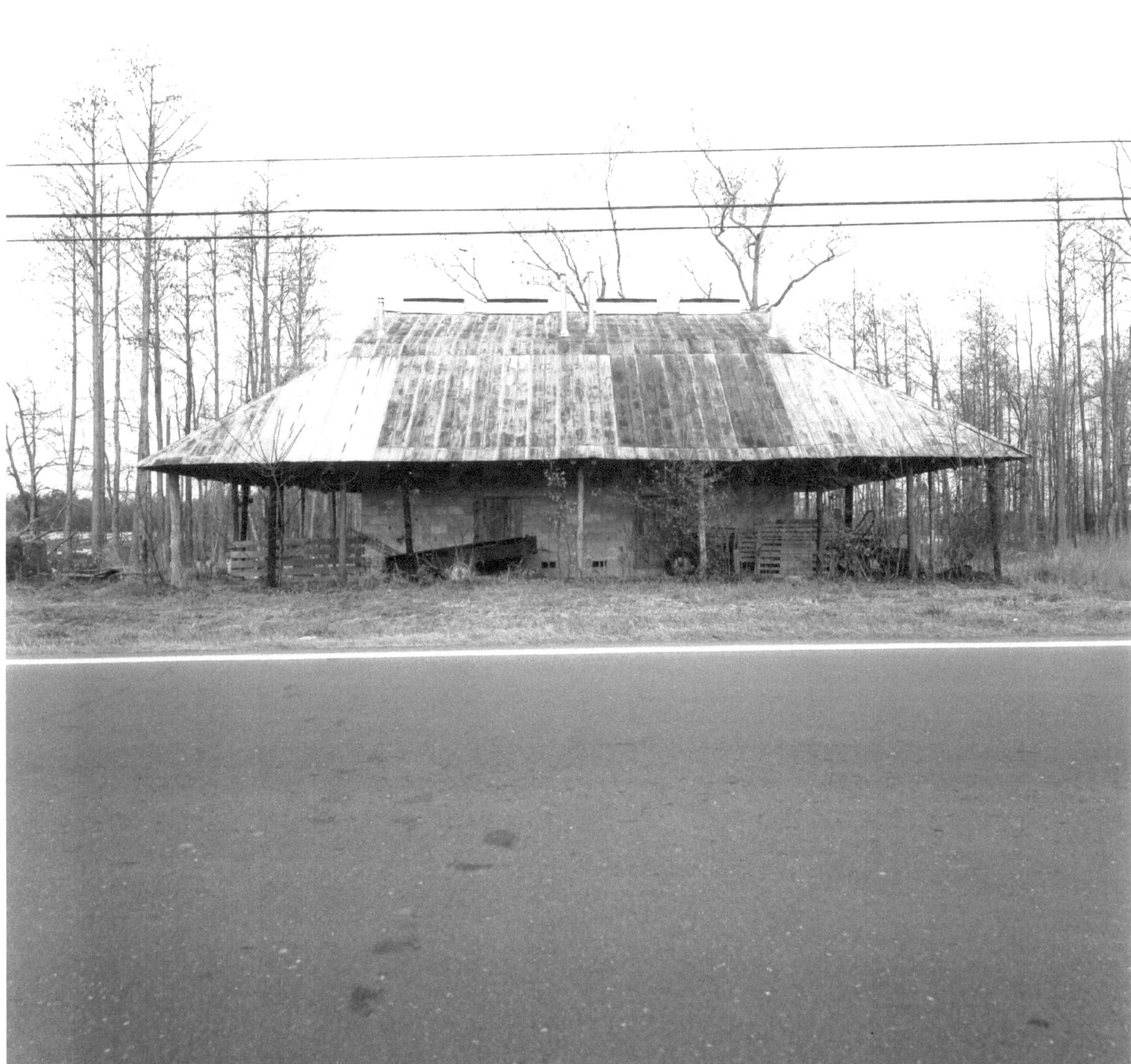

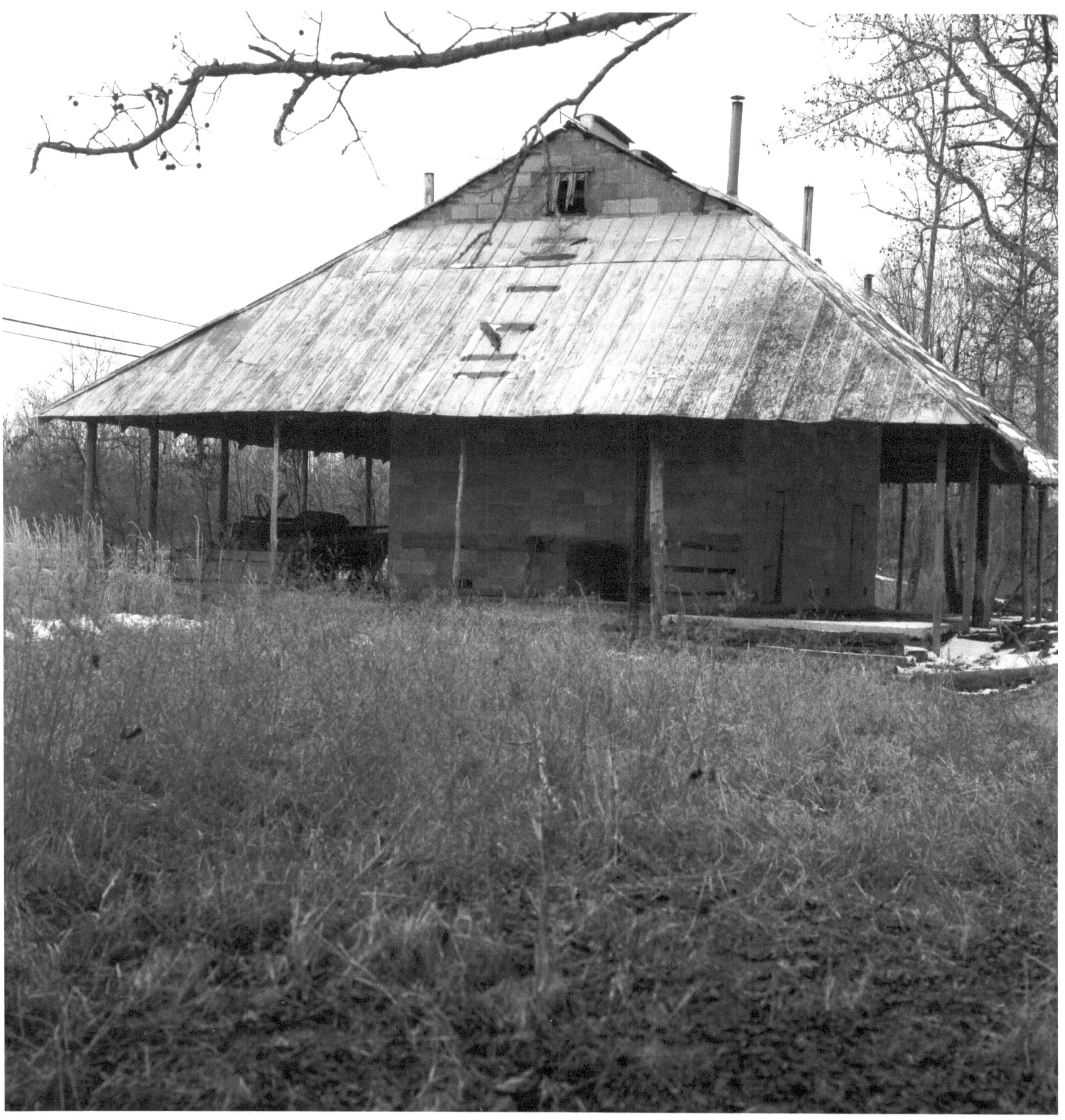

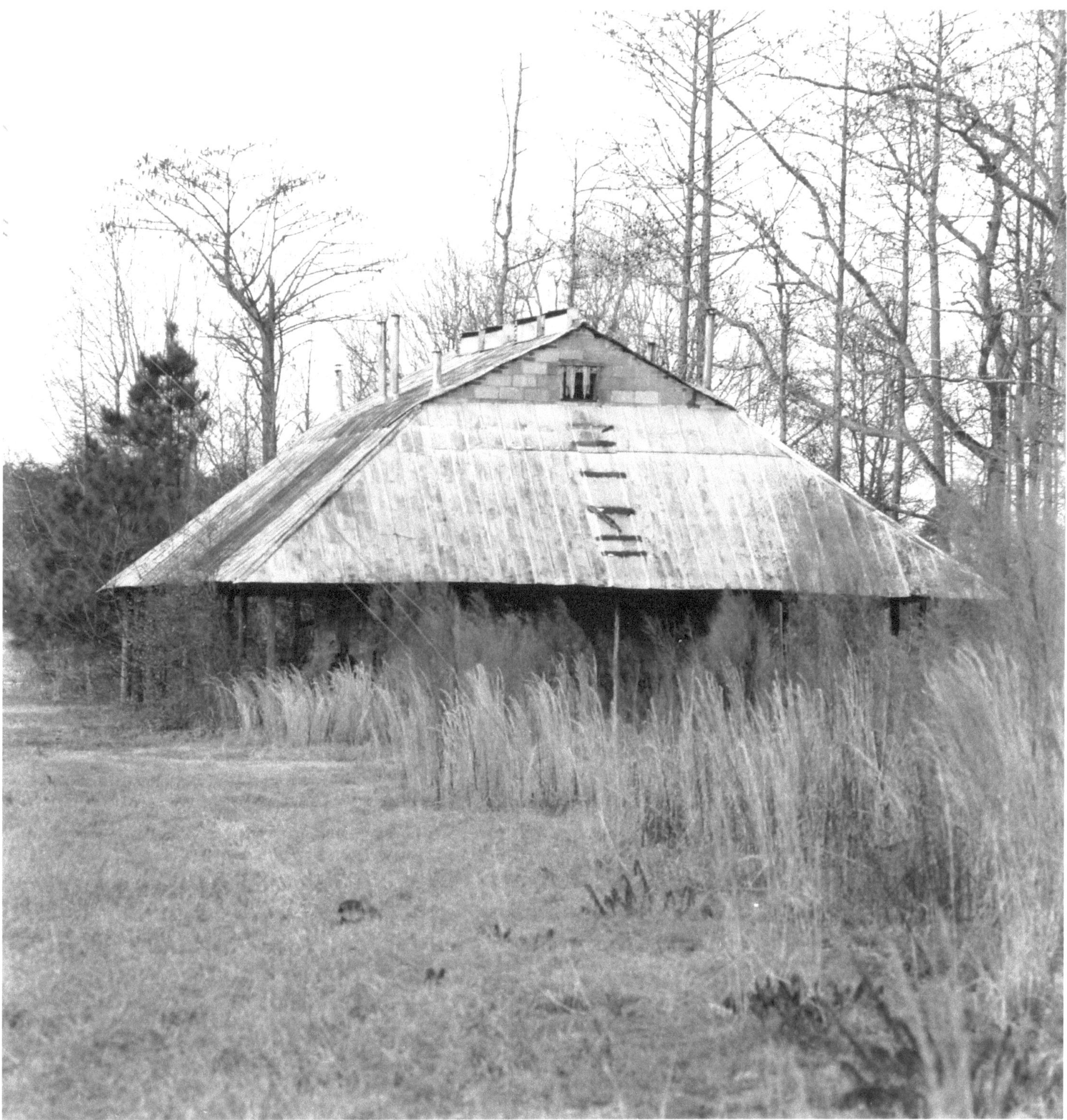

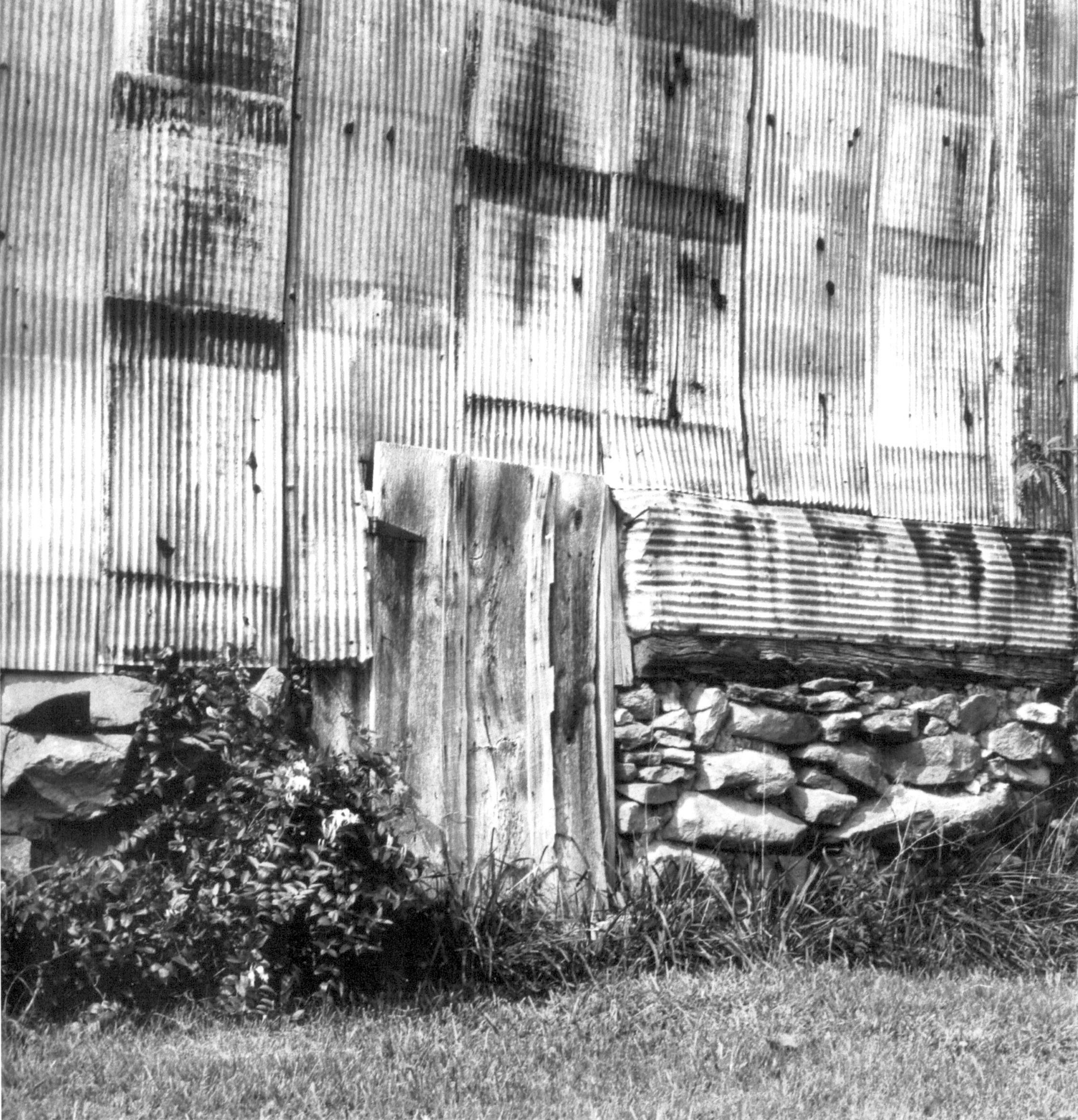

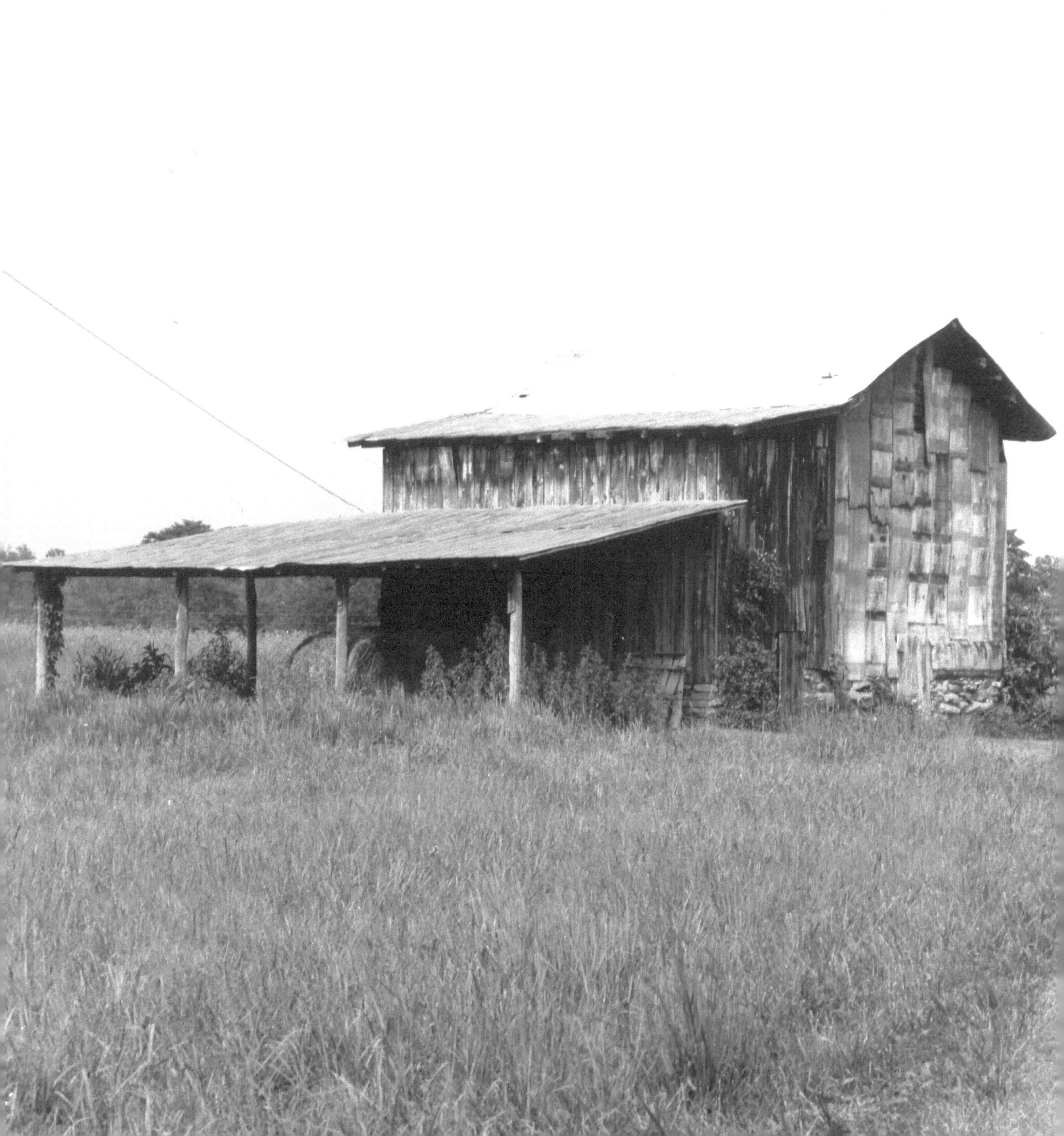

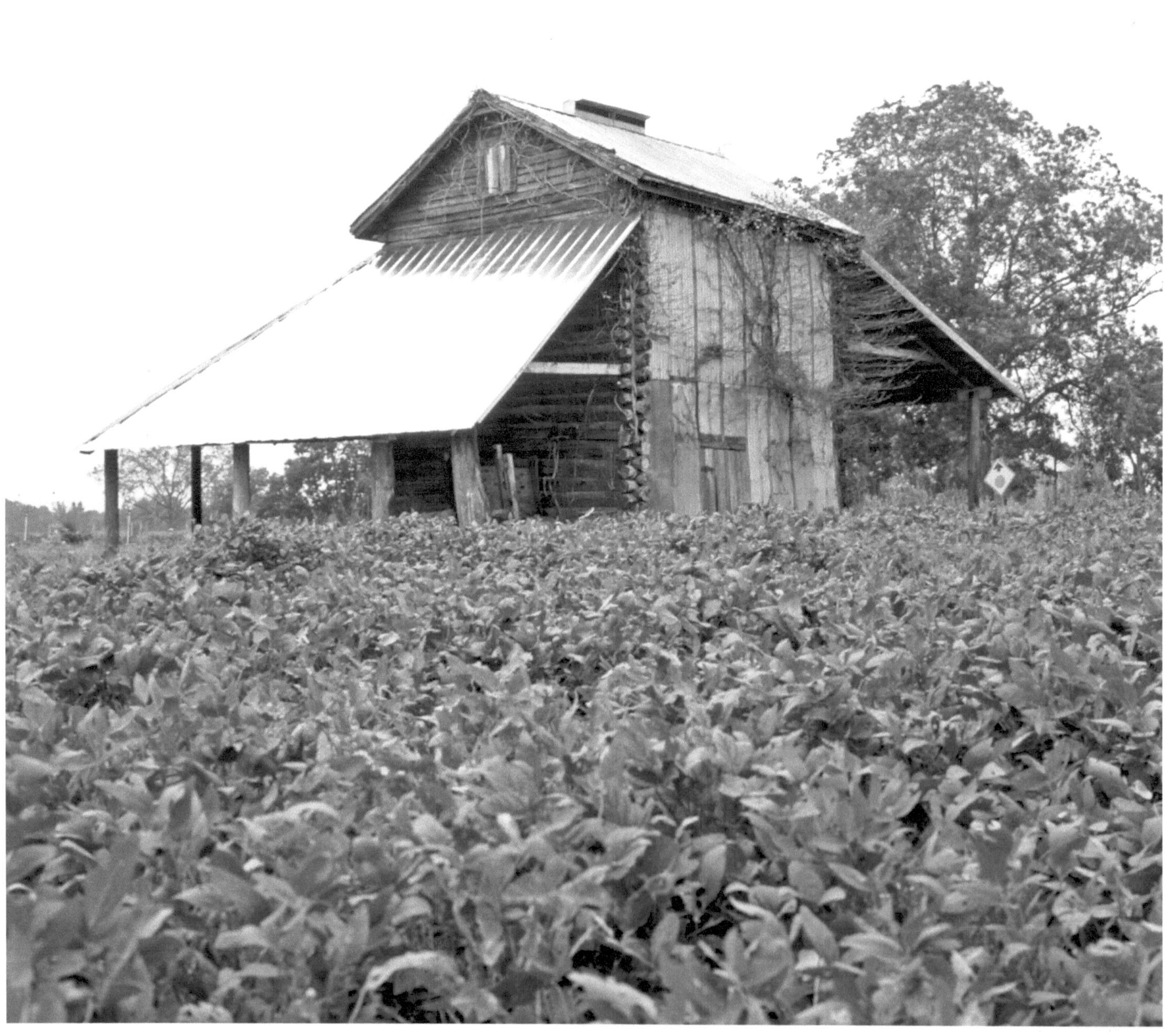

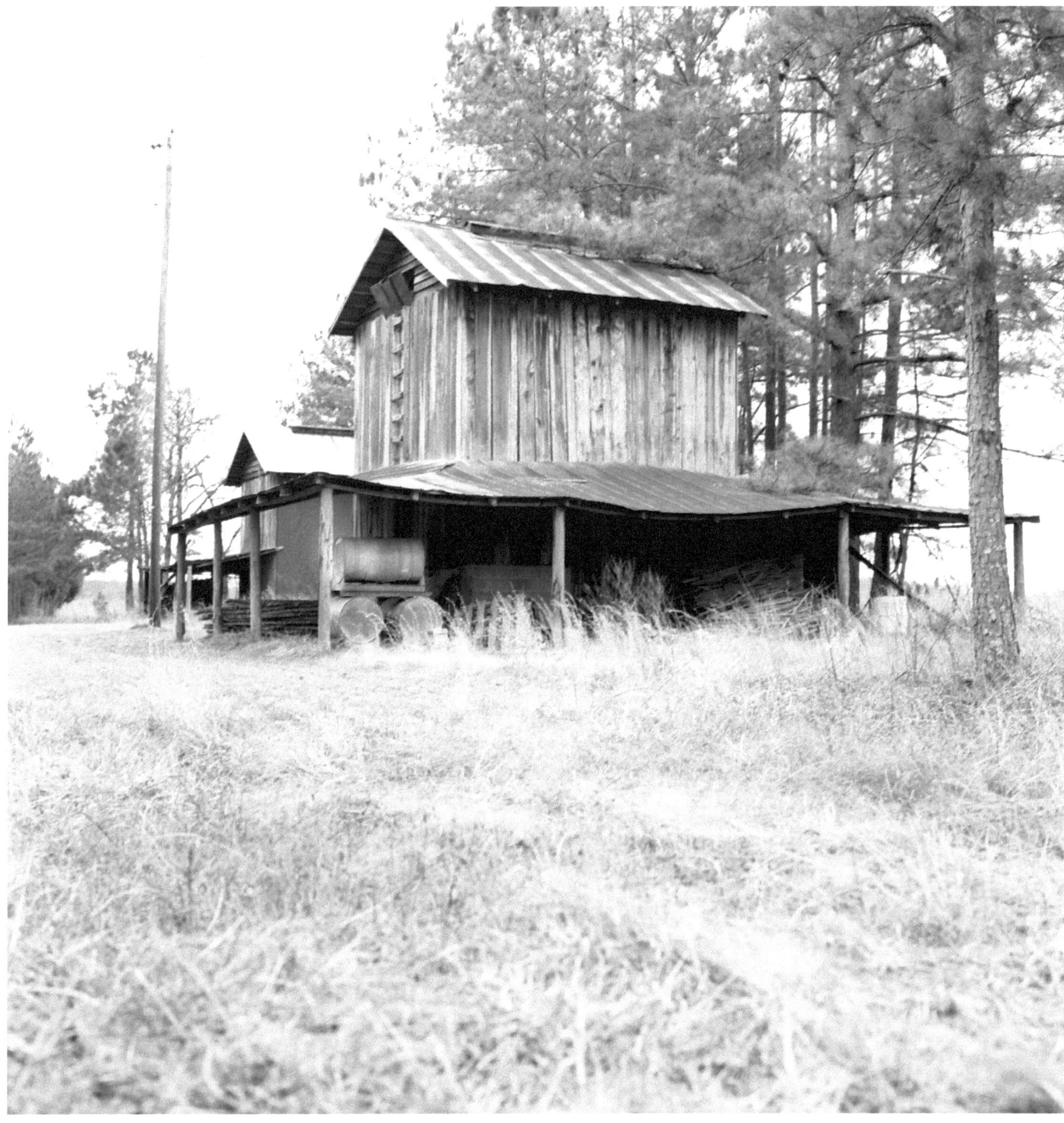

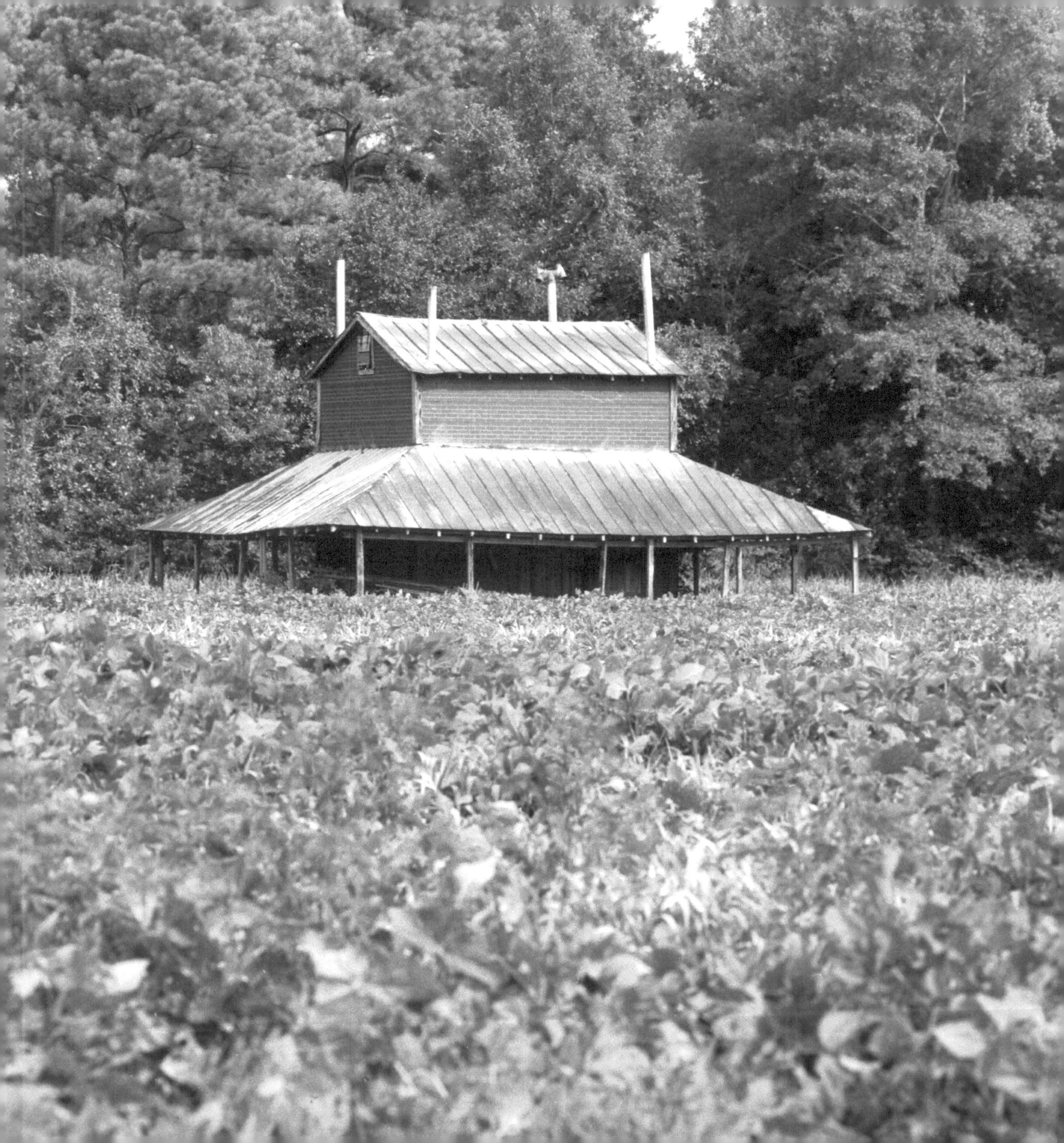

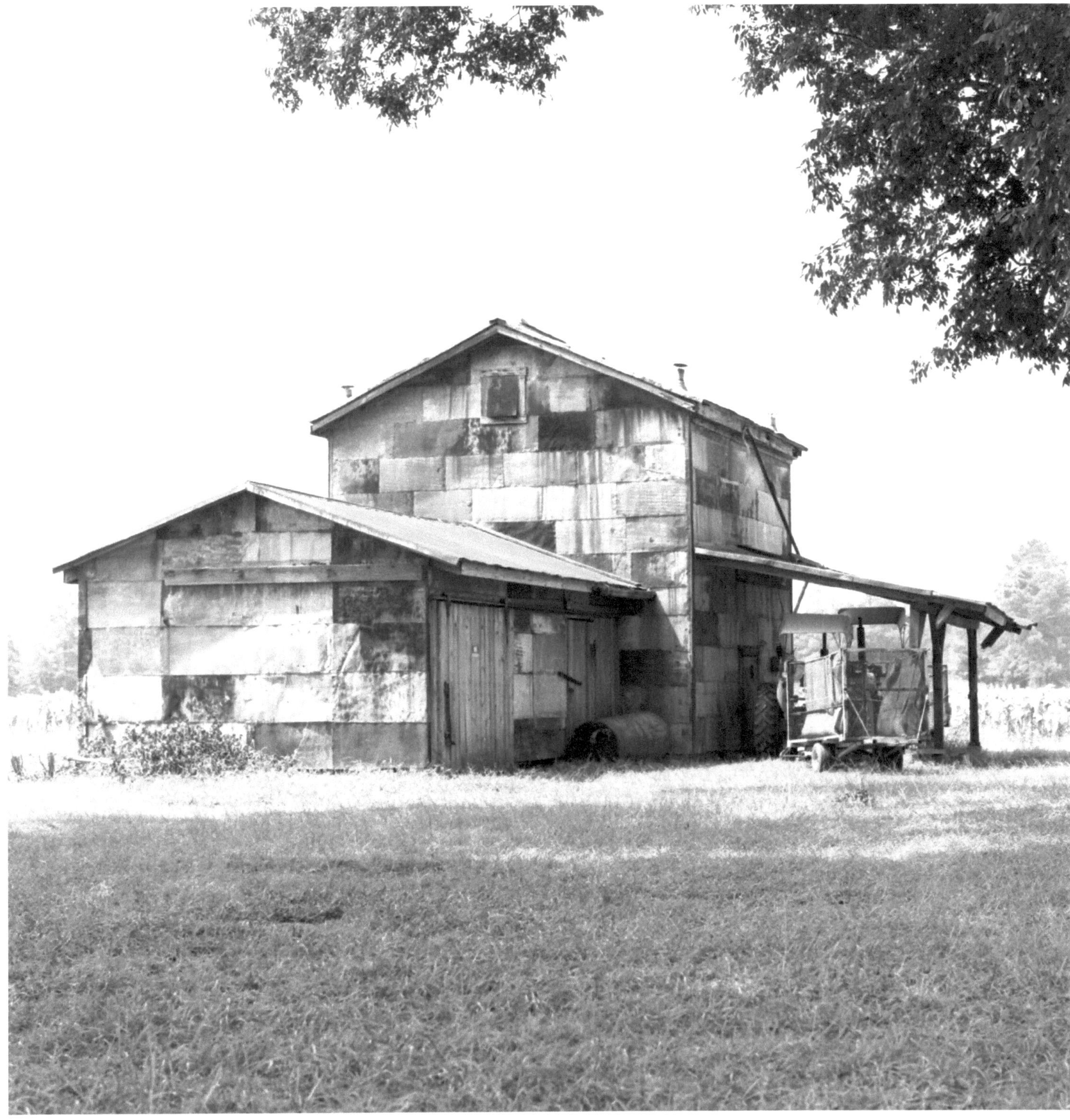

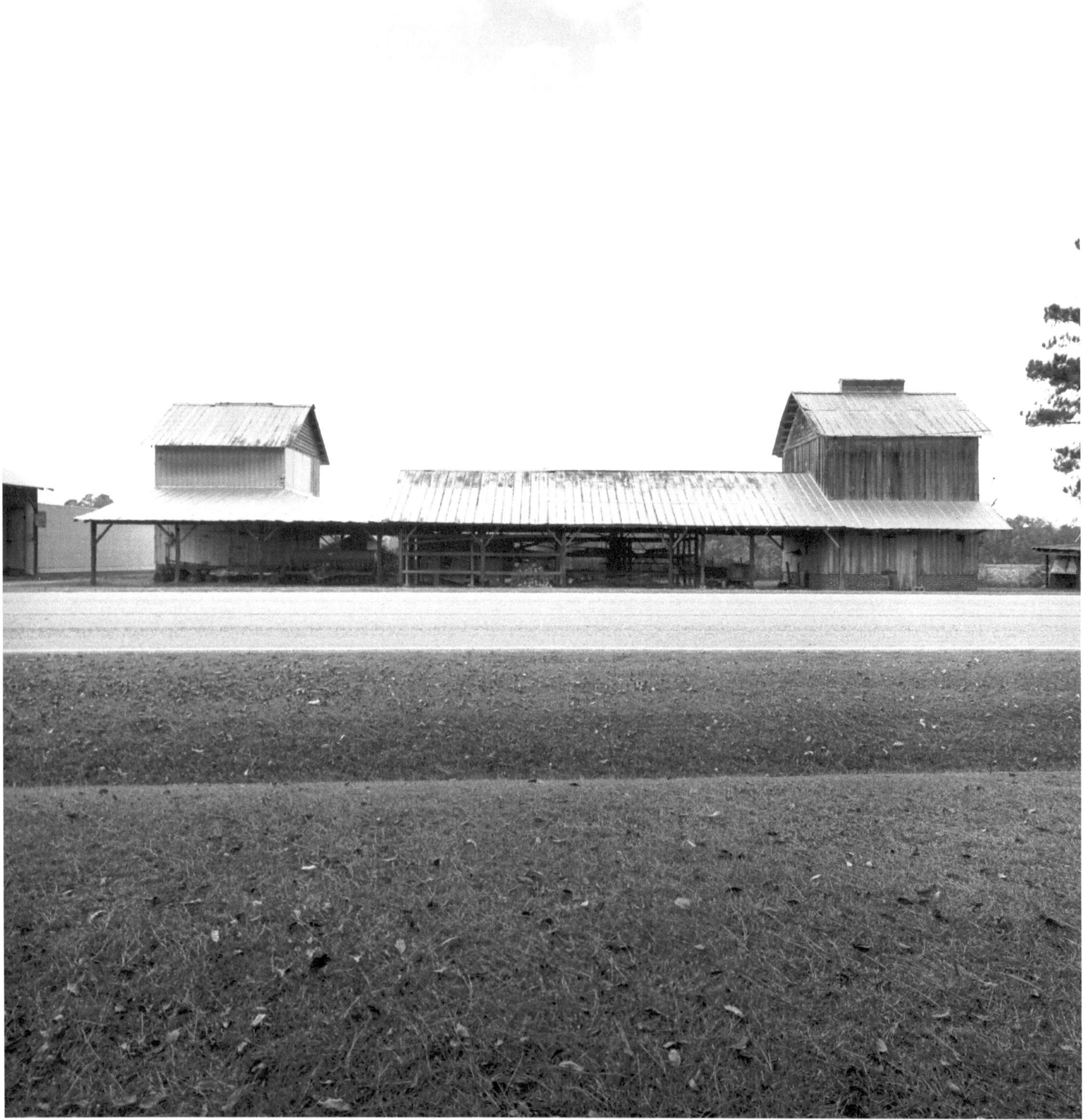

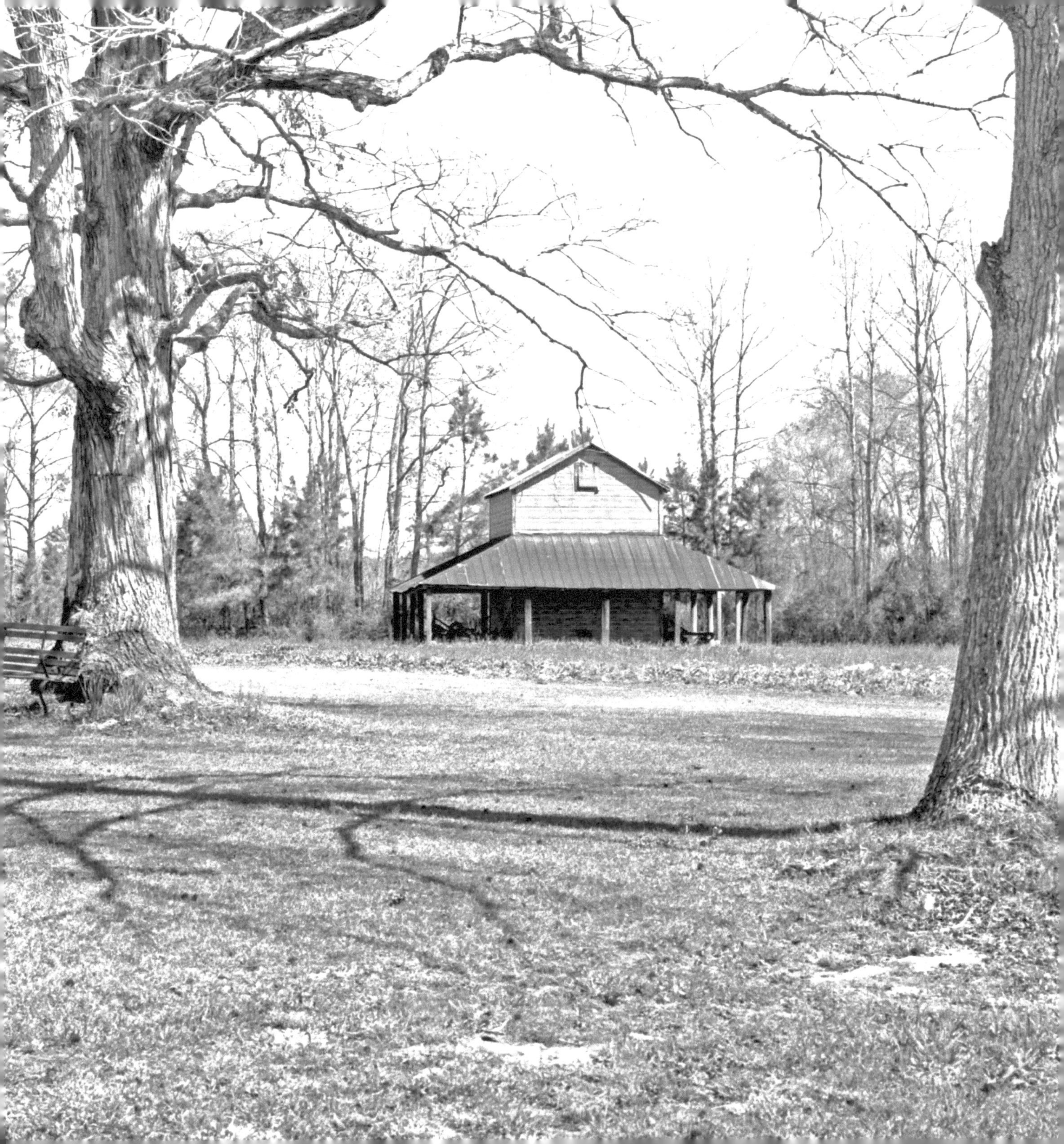

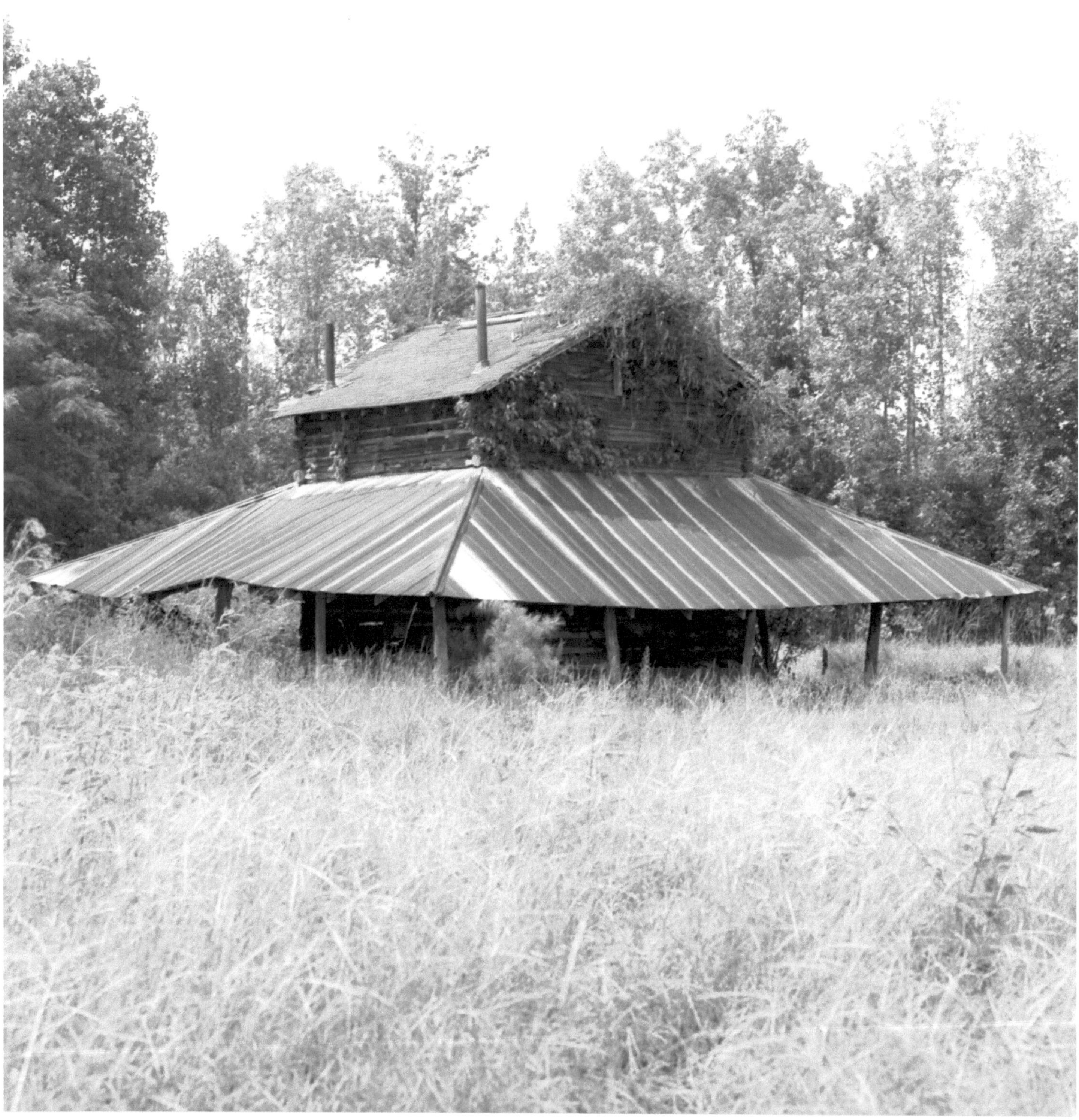

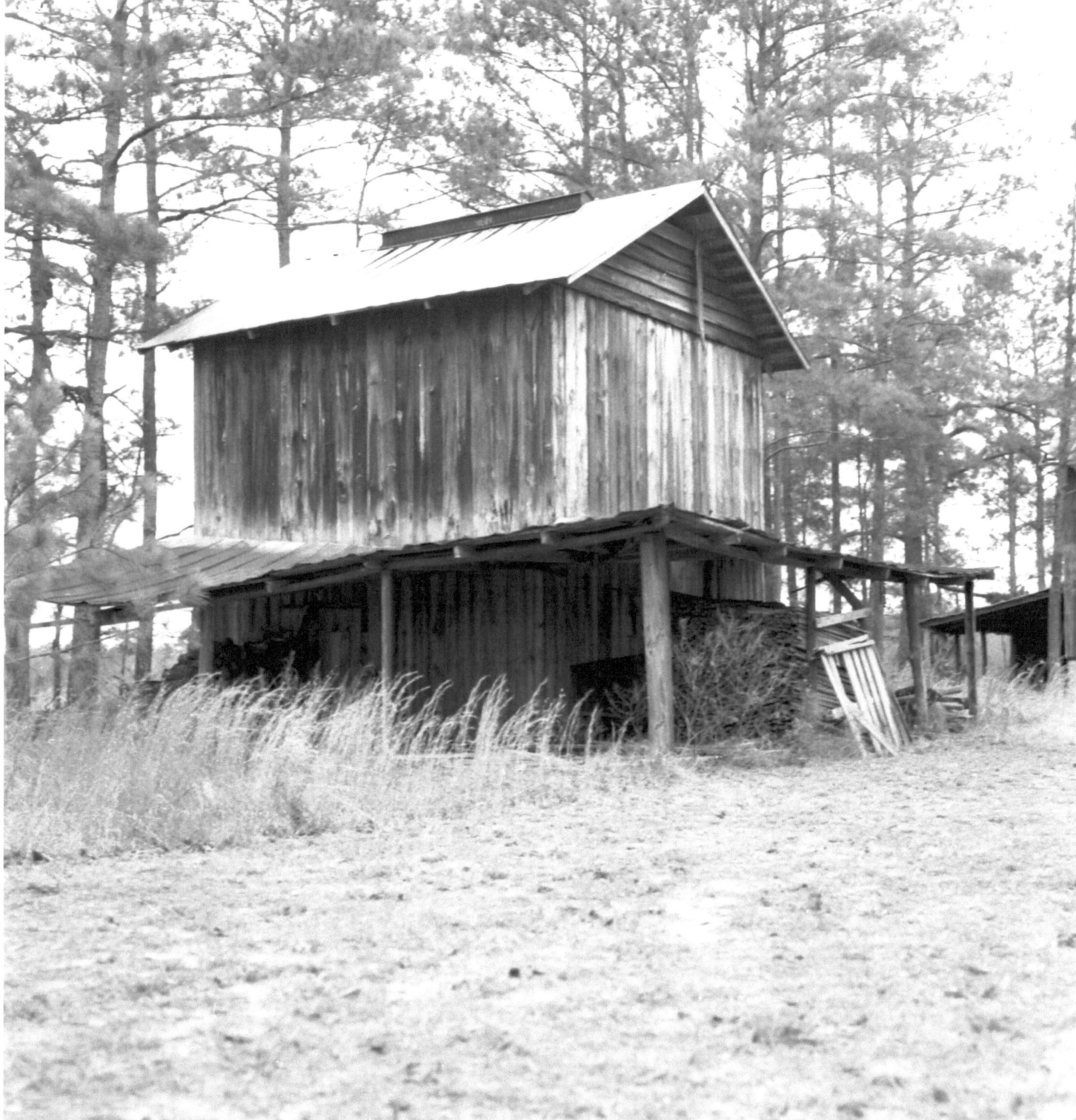

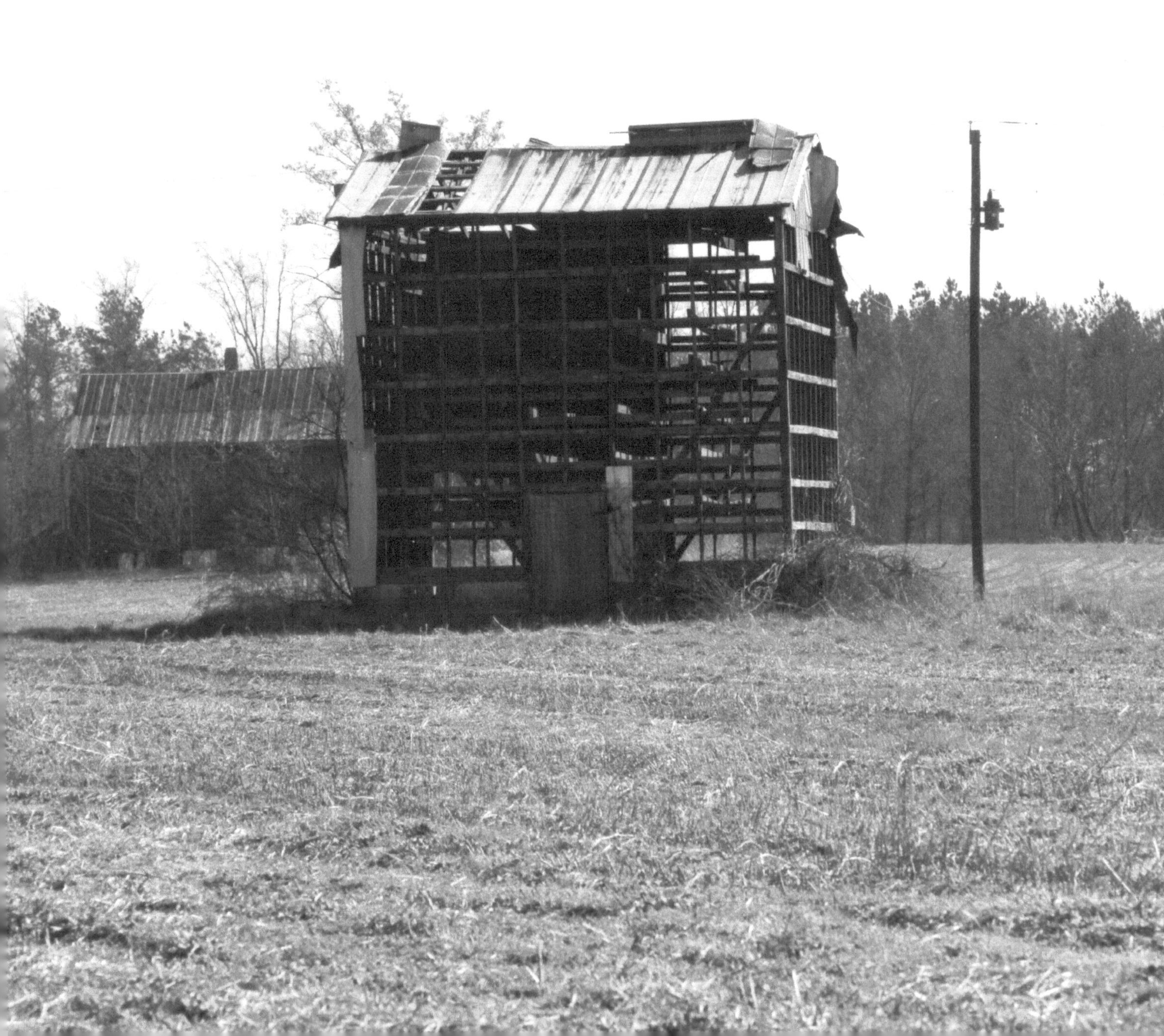

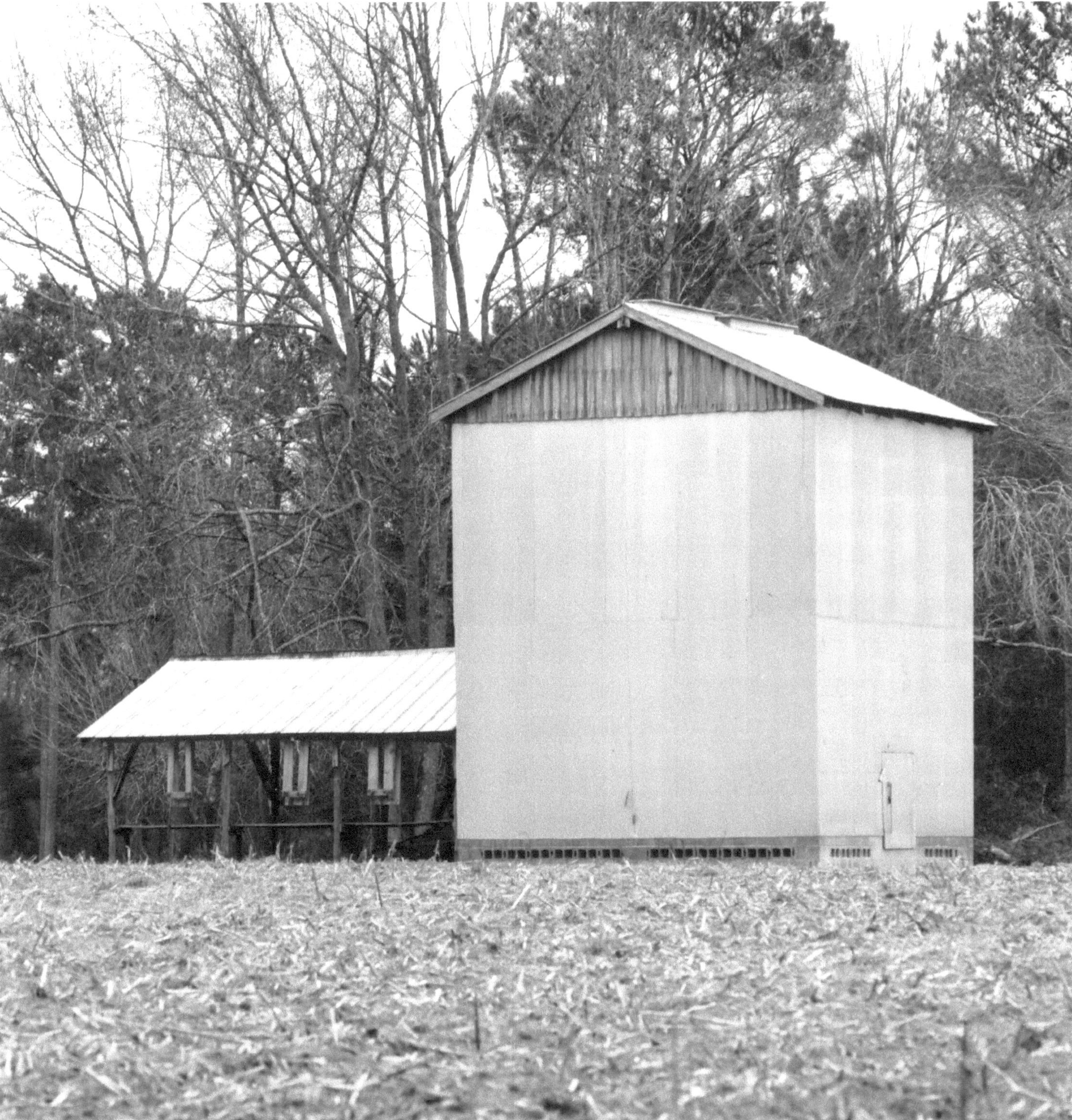

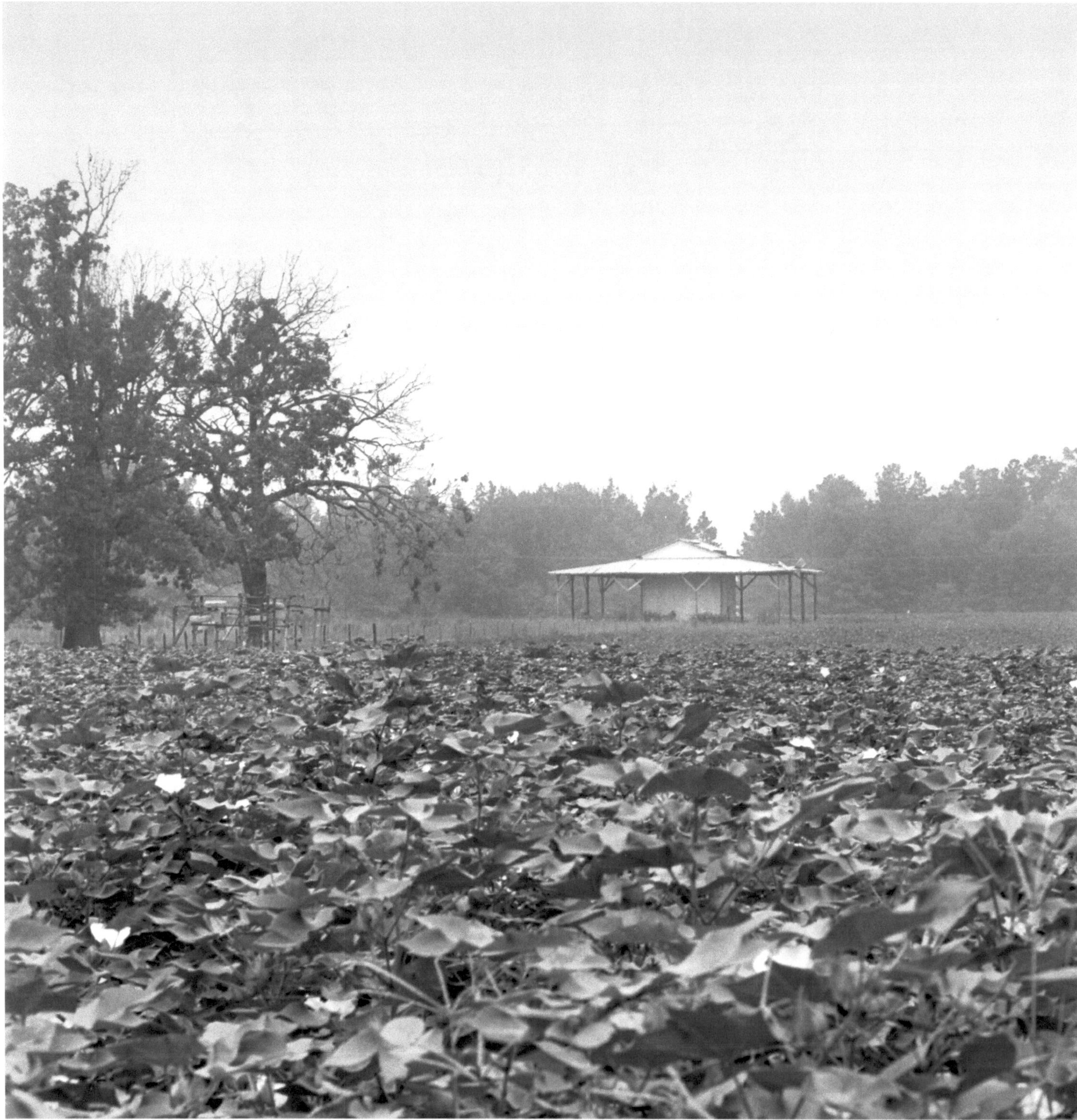

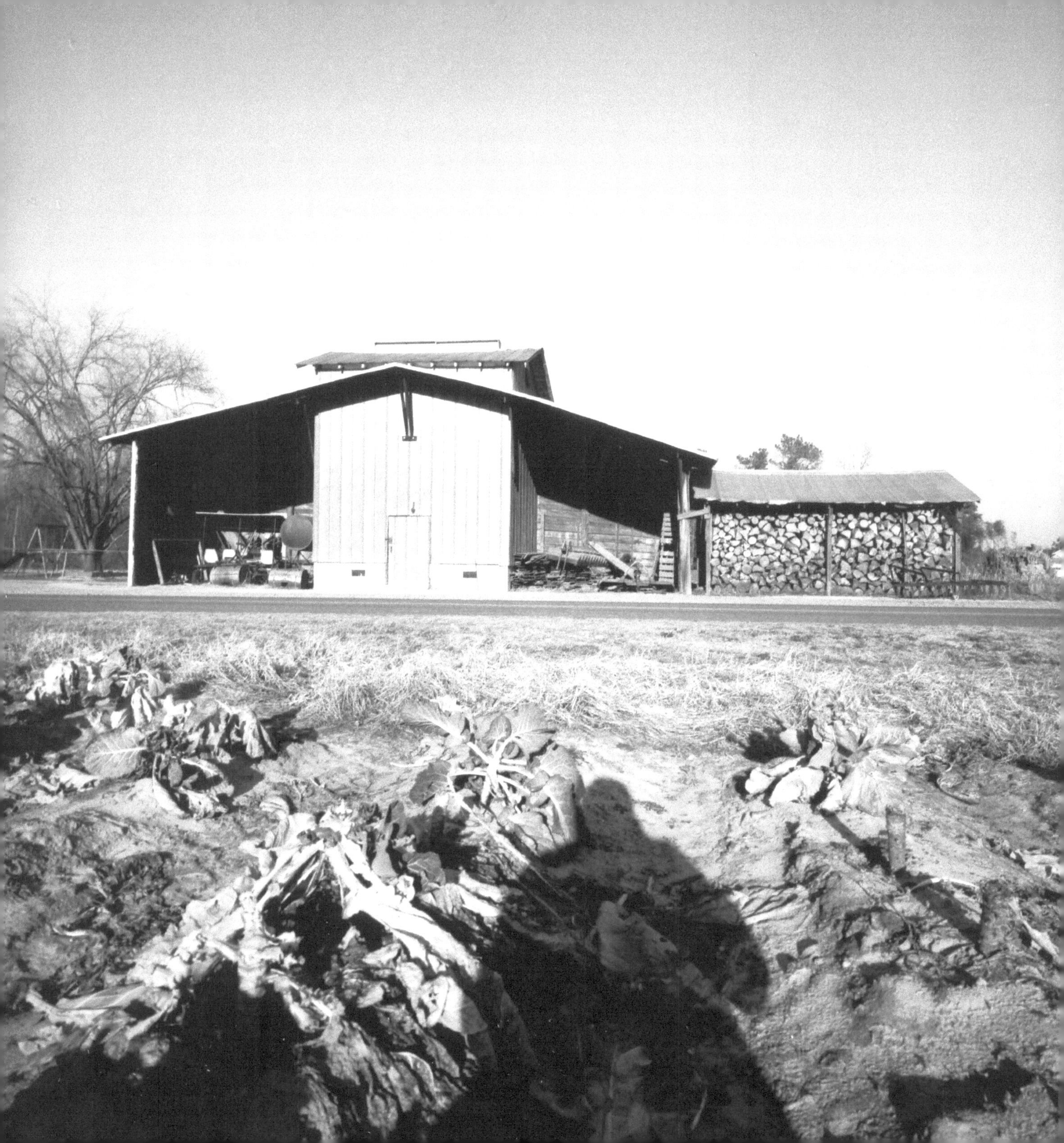

www.ingramcontent.com/pod-product-compliance
Lightning Source LLC
LaVergne TN
LVHW070133110826
845147LV00002B/243

* 9 7 8 0 6 1 5 1 3 9 8 1 4 *